This is the Flow

This is the Flow
The Museum as a Space for Ideas

Edited by Rutger Wolfson

With essays by:
Guus Beumer
Cornel Bierens
Valentijn Byvanck
Edwin Carels
Chris Darke
Bregtje van der Haak
Bas Heijne
Anna Tilroe
Rutger Wolfson

Valiz, Amsterdam

9   Foreword

11   Rutger Wolfson:
This is the Flow
The Museum as a Space for Ideas

51   Anna Tilroe:
The Promise

59   Bas Heijne:
Reality

85   Cornel Bierens:
We'll slide down

95   Bas Heijne:
Real Seeing

107 Bregtje van der Haak:
Interview with Tim Stoner

115 Rutger Wolfson:
Art in Crisis

131 Valentijn Byvanck & Rutger Wolfson:
Museums and Contemporary Language
Interview with Geert Mul

145 Anna Tilroe:
The Better World

153 Guus Beumer:
The Language of Fashion

183 Chris Darke:
Sublime Bodies
On the Work of Chris Cunningham

193 Edwin Carels:
Unguarded Moments

219 Anna Tilroe:
On Values and Symbols

251 Contributors

253 Credits

257 Sources

Foreword

From the difference between nightclubs and museums to the (im-)possible ambitions of art and the fading distinction between high and low culture. From the response to the increasing artificiality of reality in contemporary painting to the carefully constructed ambience of orderliness and luxury in airports. From fashion's power to seduce to its visual conventions colonizing other disciplines. From the leisure industry to the supposed innocence projected onto children by adults. From the relationship between technology and the sublime to the question of whether art can develop symbols that express the values that connect us today. The highly divergent subjects in this collection of essays were the inspiration for an exhibition, or vice versa: the ideas in an exhibition were the inspiration for an essay. The exhibitions in question were all held at De Vleeshal – a centre for contemporary art in Middelburg, the Netherlands, where I served as director from 2000 to mid-2008.

As I worked on these exhibitions, I gradually became conscious that I was wrestling with an underlying question, one that connected these highly varied projects: what role do the visual arts and museums play in our society? And perhaps more importantly, what role *might* they play?

The introduction to this volume is an attempt to answer that question. Museums must re-establish their legitimacy and engage in a more explicit relationship with society, I argue. How they achieve this is up to each museum to figure out for itself, precisely because this is a question of its specific individual identity.

This is the reason I decided to base this volume of essays on a series of exhibitions at De Vleeshal. In doing so I wish to illustrate how a museum can reflect on society using highly diverse subjects and from highly different angles. Of course De Vleeshal is hardly the only institution to choose this sort of approach, and the issue of the legitimacy of museums is frequently examined elsewhere. By approaching this issue from a specific place, however, I hope to make a specific contribution to the debate on the future of the museum.

Because of the exchange of ideas from which it emerged, this book is a collaborative effort in more ways than one. I am, therefore, extremely thankful to all the authors in this collection and to all the people who worked on the exhibitions at De Vleeshal. Most especially to my colleagues – without them, these ideas would have remained just that.

Rutger Wolfson, May 2008

Rutger Wolfson:
This is the Flow. The Museum as a Space for Ideas

1. Art and Reality

Sunk back in the rear seat of a taxi on the way
from the airport to my hotel, I try to take in
as much of Sao Paulo's high-rise buildings as
possible. The view reminds me of the science-
fiction classic *Blade Runner*.[1] The next morning,
suffering from jetlag, I set off in search of several
galleries, but have little idea of the way. By chance
I stumble upon the Museo do Esculpture. The part
of the museum situated above ground – a sort of
sculpture garden – passes me by in a haze. Looking
for the toilets, I walk down a ramp. Beneath it
hangs the kind of circular mirror used to reflect
oncoming traffic. Initially I think I have walked
into a car park and I stop, somewhat confused. As
I turn the corner I become completely disoriented.
Positioned in a large subterranean space are
luxurious, white leather seats around stainless-
steel poles, a turnstile, stainless-steel handrails
and an enormous mirrored wall – all brightly lit.
At first I think I am in the museum's offices, a split
second later in a recently opened metro station.
I am perplexed by the staged, yet hyper-realistic
situation in which I find myself. A full minute
passes before I convince myself that this must be
a work of art.

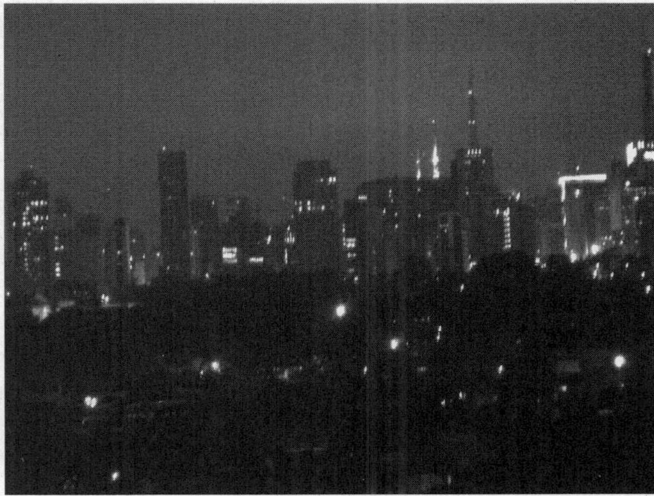

1.1   São Paulo at night, 2005

[1] Ridley Scott, *Blade Runner*, 1982.

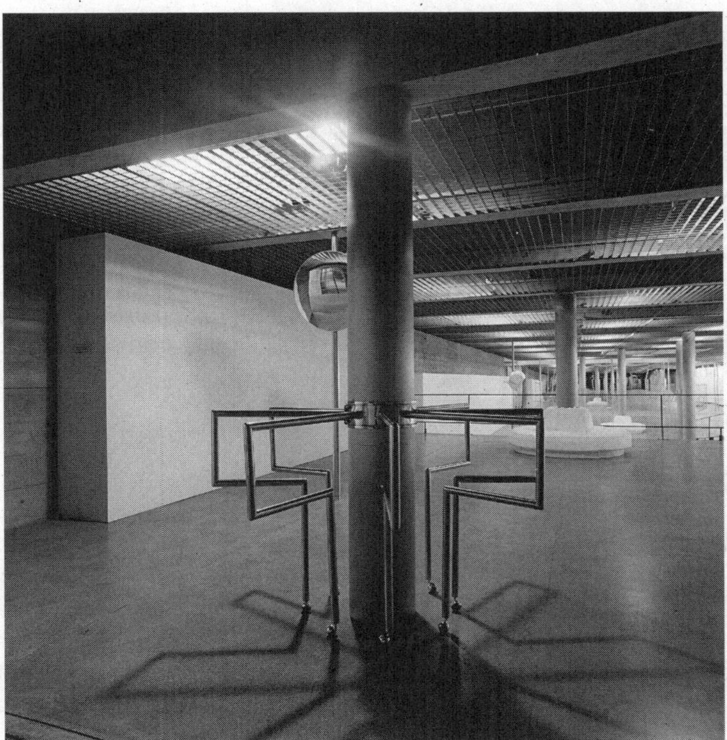

1.2   Ana Maria Tavares, *Relax'o'visions*, 1988

Ana Maria Tavares, the artist responsible for this installation, is fascinated by so-called 'non-places'. Non-places are sites such as stations, stadiums, airports and hotel lobbies all around the world, which resemble one another so closely that they are simultaneously everywhere and nowhere. Non-places lack specific references to the cultural and historical identity of the place. Schiphol Airport could just as easily be Singapore Airport and vice versa. For this reason, non-places are often associated with a feeling of detachment. In her article in this volume, Anna Tilroe outlines how Tavares, at an initial level, appears to be attracted to the seductive nature of these non-places. In the work Middelburg Airport Lounge with Parede Niemeyer she explains: 'Everything is under control and, at the same time, full of promise: we have left, but we have not yet arrived. Things have yet to happen; time, for the moment, is suspended' (see pages 51-57). A pleasant, detached intoxication. But at a deeper level, Tavares clearly demonstrates what non-places attempt to hide: control over the people that use them. The stainless-steel handrails, whose primary function is simply to provide support, also appear to direct our movements. The turnstile is designed both to afford us access and to deny it.

My somewhat disoriented state when I first saw the work of Ana Maria Tavares must no

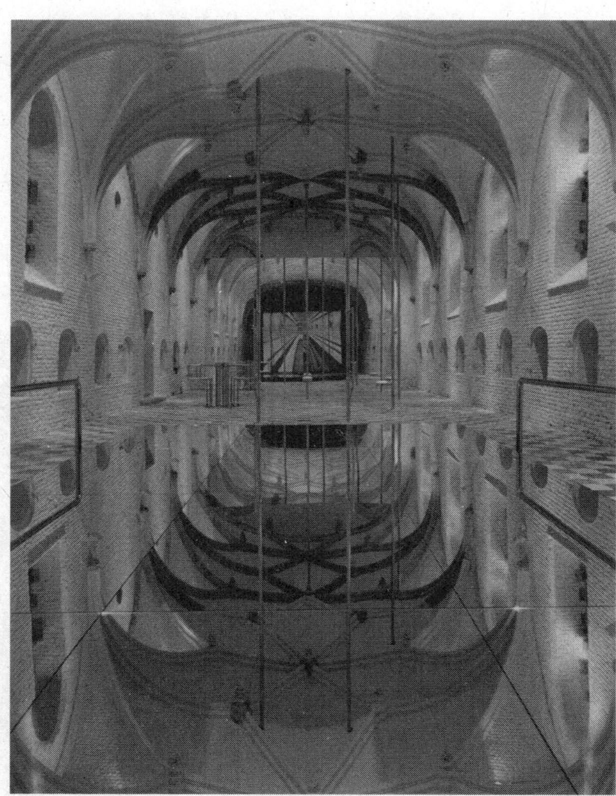

1.3   Ana Maria Tavares, *Middelburg Airport Lounge with Parede Niemeyer*, 2001

doubt have played a part in my response, but the impact was nonetheless enormous: a euphoric moment in which a long-slumbering insight suddenly became clear. Since my encounter with Tavares's work, I have an altered regard for airports and all other kinds of non-places. I can now appreciate the finesse and intelligence with which a non-place has been fitted out. More than before, I am aware that my stay is designed according to a strict *mise-en-scène*. Tavares has, as it were, made non-places legible and thus bearable. In this way Tavares's work illustrates an important characteristic of art, namely its ability to offer insights into ourselves and the world around us. Art has, in other words, the ability to bestow meaning.

This power is now of greater importance than ever. The world is changing rapidly and becoming ever more complex. Tavares's work shows us that modern reality is becoming increasingly artificial. In the design of our everyday environments – in the landscape, shops, in the public realm and at home – spectacle and experience are often more important than authenticity. But this modern reality does not manifests itself only in the physical world. The fact that artificiality has also become an increasingly important component of our inner world is made clear by comparing the characters from two novels.

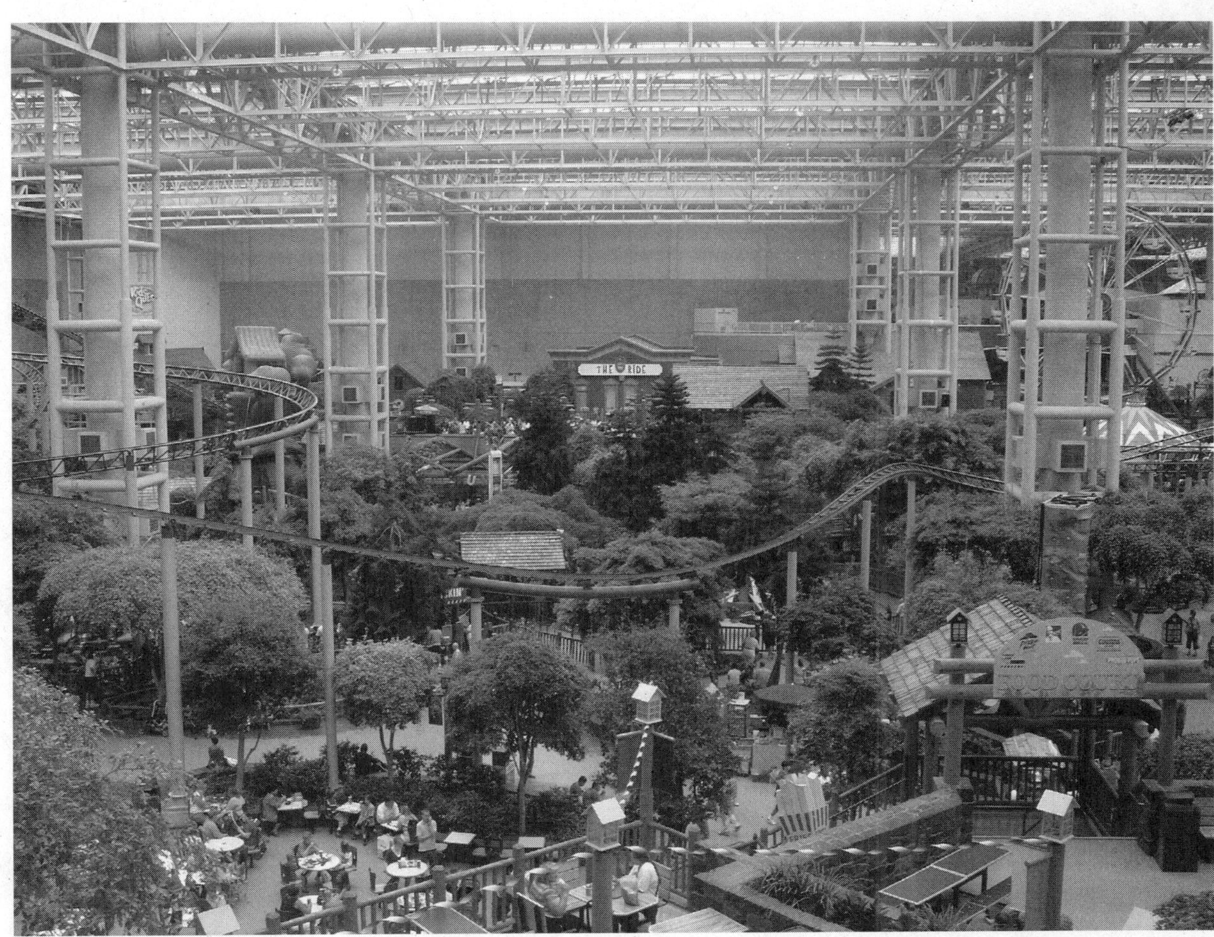

1.4   Mall of America, 2006

The hero of J.M. Coetzee's novel *Youth*, published in 2002 but set in the 1960s, is a classic romantic.[2] He has studied mathematics in South Africa but wishes to move to Europe and become a poet. He has no clear strategy; he knows only that he wants to live in London. He would actually prefer to live in Paris – the city of poets and artists – but he does not speak the language. Upon arriving in London he finds a squalid flat and gets a job as a computer programmer.

He dreams of an inspiring artistic milieu and of impassioned European women. But the life he seeks appears unattainable. His lack of human contact makes his life a martyrdom, but he consoles himself with the idea that suffering nurtures art. His head is full of the poet he wishes to become, but the writing itself is increasingly difficult. He roams the streets in search of a woman who will return his gaze and understand him instantly. He hankers after a love that, in a blinding flash, will make him complete, as written about by his favourite poets. A love that will empower him to write poetry and to fulfil his ambitions, his life. Towards the end of the book he is threatened with a humiliating return to South Africa. He is destitute, he has failed to find love and his attempts at writing have come to naught. Reluctantly he begins to wonder whether things would not have been better had he taken the

[2] J.M. Coetzee, *Youth*, Secker & Warburg, London, 2002.

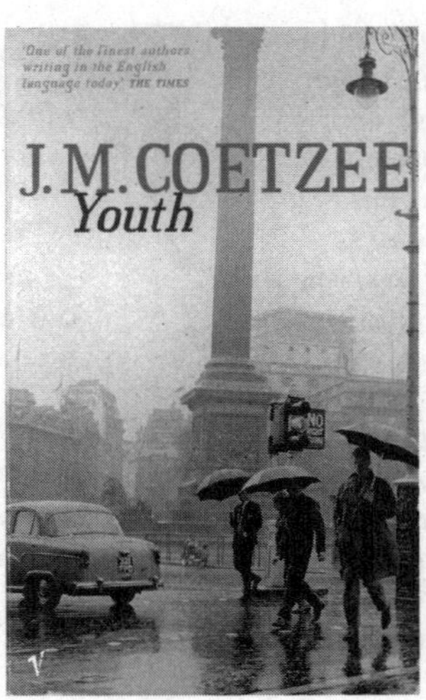

initiative himself rather than longing for love to reveal itself to him.

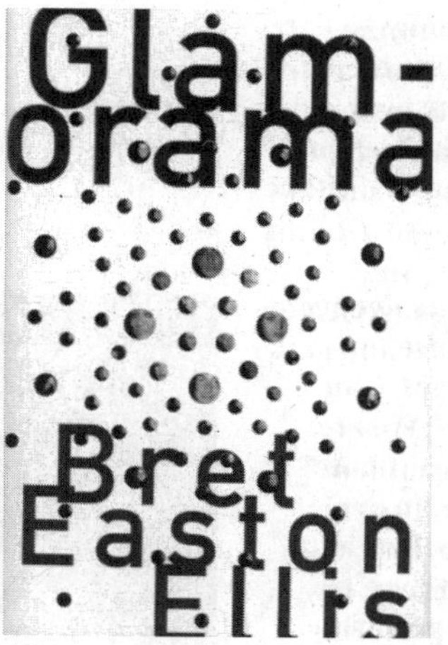

At first glance, it is hard to imagine a greater contrast between the hero of Coetzee's novel and Victor Ward, the main character of Bret Easton Ellis's 1998 novel *Glamorama*.[3] Ward is a successful model in New York at the beginning of the 1990s. In *Glamorama*, the romantic longings of *Youth* have made way for narcissism. Ward wishes to see only his own image, either in the eyes of others or in the media. His world is dominated by what is *en vogue*. Everything is superficial, or at the very least has a gloss of irony and ambiguity. He is committed to nothing. At most, when the situation requires it, he feigns a faux commitment.

There is a world of difference between the modern reality that Ellis describes and that of the 1960s in Coetzee's book. Most conspicuously, there is little or no sign of the influence of mass culture in Coetzee's novel, whereas in *Glamorama* (set some 30 years later) it is ubiquitous. Ward's inner world, inasmuch as it exists, is saturated with images from the mass media, which colour and condition his feelings and desires. He no longer has authentic experiences; for him a situation or emotion becomes real only if it reminds him of an image from a newspaper, music video or film. When anyone asks him how he feels he replies

[3] Bret Easton Ellis, *Glamorama*, Alfred A. Knopf Inc., New York, 1998.

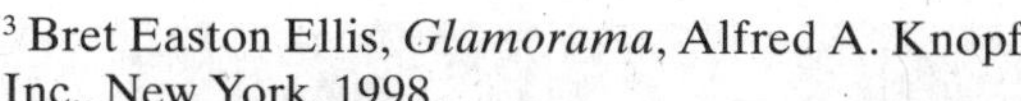

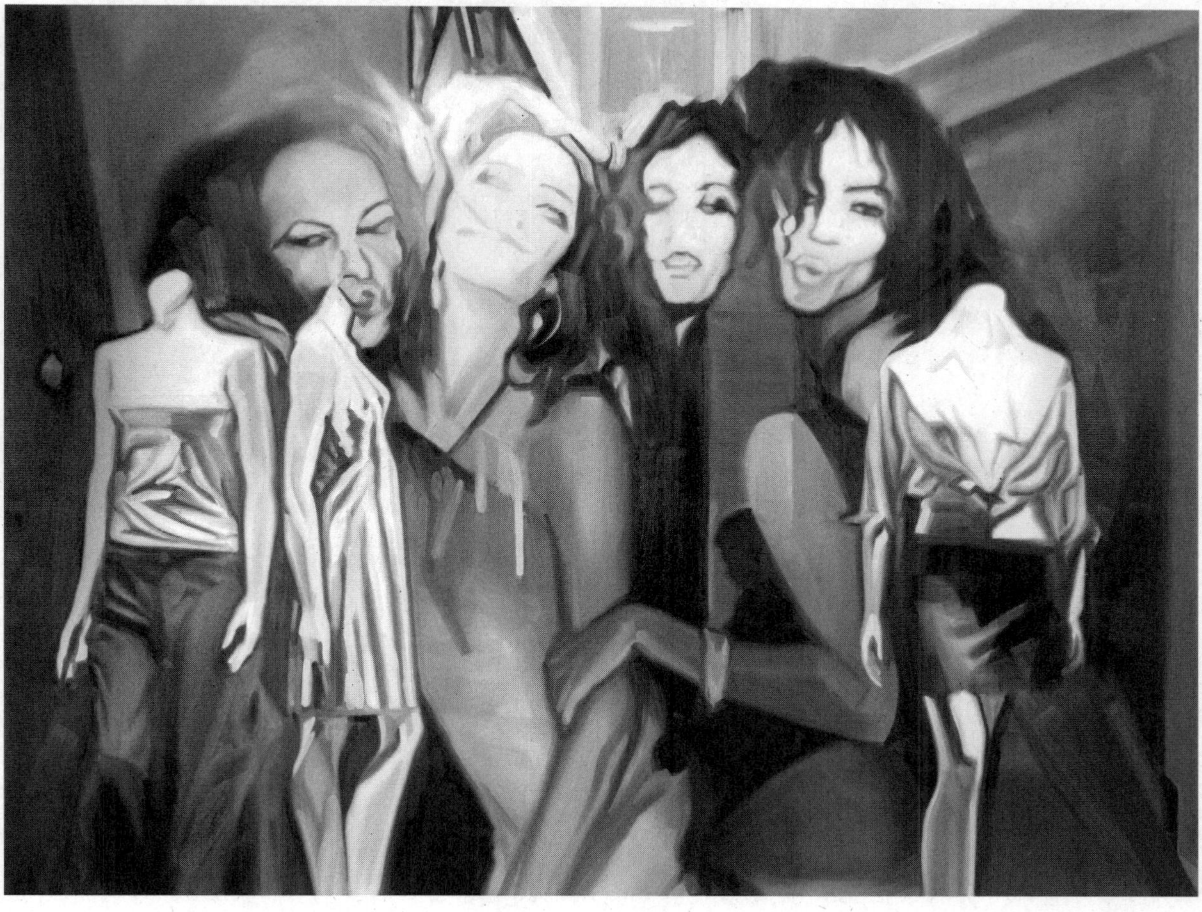

1.5   Cameron Rudd, *West Witches*, 2003

with quotations from song lyrics. He longs to live in the artificial world evoked by the mass media. As the book proceeds and his consciousness gradually disintegrates further, he deludes himself that he is being filmed in a sequence of sets, either by chance or by design.

Now that the mass media penetrate deeper into our lives, the question of how we are to relate to reality has become more urgent. Can art provide an answer to this question? We are inclined to see art and museums as an antidote to the mass media's harmful influence on our reality. Art and museums stand for civilization, profundity and import. In this respect they are the diametric opposite of all that is superficial and artificial. But in his essay 'Reality' (see pages 59-84) Bas Heijne goes a step further. He wonders whether art is capable (or indeed desirous) of anything more than simply existing as an opposite of superficiality. This question is, I believe, a well-formulated one. The notion that art and museums are an antidote for modern reality is too limited. Art and museums should not shut themselves off from this reality; they should attempt to offer us insights into how we might relate to it.

An example of this is the exhibition *We'll slide down the surface of things*. The title is borrowed from a party scene in *Glamorama*

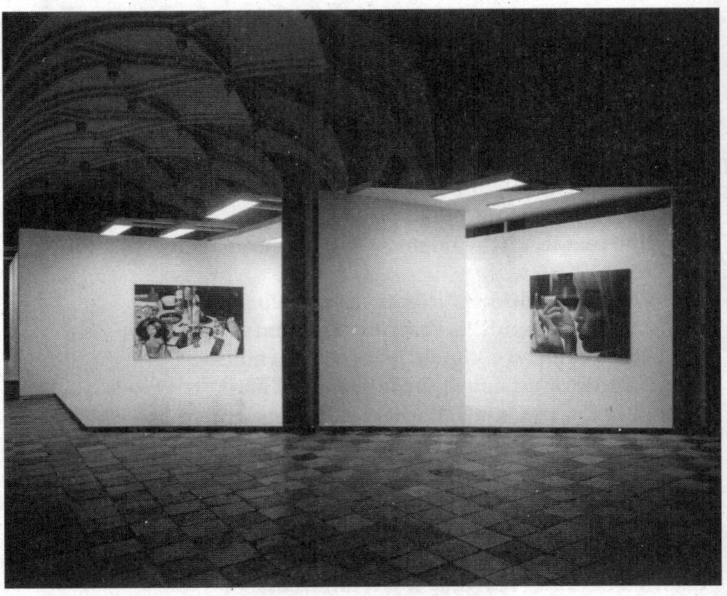

1.6   *We'll slide down the surface of things…*, installation view, 2002

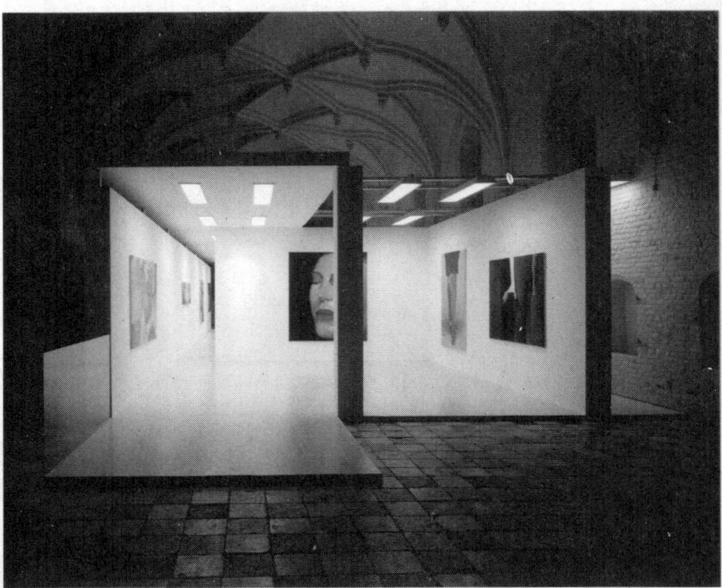

1.7   *We'll slide down the surface of things…*, installation view, 2002

in which Ward repeatedly, and with a mock-profundity, quotes these lyrics from a U2 song. The exhibition used paintings to explore how we can relate to our increasingly artificial reality. 'To condemn this is pointless; it would be better to accept the predilection for appearances as a given and, in so doing, to explore whether new meanings can nevertheless be attached', writes Cornel Bierens in his essay about this exhibition (see pages 85-93). 'Reality, in all its ostentation, has become so obtrusive that painting has no other recourse but to grab the bull by the horns and face it head on. Recording superficiality straight out: that's the way *to slide down the surface of things.*'

Each of the three painters in this exhibition does this in his own particular fashion. On the face of it, Frank Bauer's paintings could be scenes from the work of Bret Easton Ellis: he portrays his friends in clubs, at a house party and at the end of a long night out. Bauer's paintings exude a longing for glamour. He illustrates our desire to see moments from our own lives from a distance, as in the stylish images we see in the mass media. Arnout Killian fits within a tradition of painters who are fascinated by light. But whereas other artists in this tradition have mainly attempted to represent natural light in landscapes or still lifes, Killian is principally interested in artificial light: he paints the hard, contrasty light of catwalks

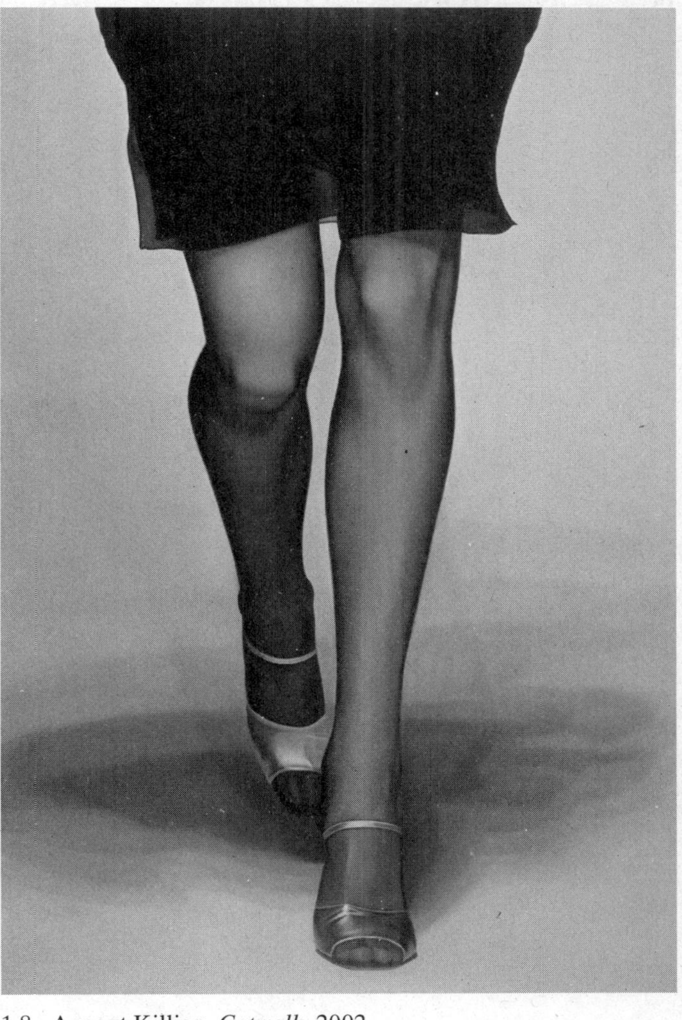

1.8   Arnout Killian, *Catwalk*, 2002

and shop windows or the cool blue light of images on the internet. This renders Killian's paintings simultaneously traditional and contemporary. Glen Rubsamen's work is equally rooted in tradition yet entirely modern. His landscapes combine a lyrical regard for nature's beauty with the uneasy emptiness of a motorway lay-by. The romantic ideal of unspoiled nature is replaced in Rubsamen's paintings by an image of nature propagated by holiday brochures and Euro Disney, and lampposts occupy a natural place alongside palm trees within his dramatic compositions. Beauty is not lost in these partially artificial landscapes; it has become ambivalent.

Modern reality has not only become more artificial but also less real. According to Bas Heijne the connections between our consciousness and reality have been broken and our emotions are increasingly predetermined, bringing about an unreal reality. When, for example, we witness something terrible with our own eyes, such as gratuitous violence or the attacks of 11 September, we tend to experience it as a film. We certainly have the sensation of having experienced something appalling, but we are simultaneously reduced to the position of passive bystanders. According to Heijne art should allow us to escape this unreal reality. In response to the superficiality of the mass media it is contemporary art's task to

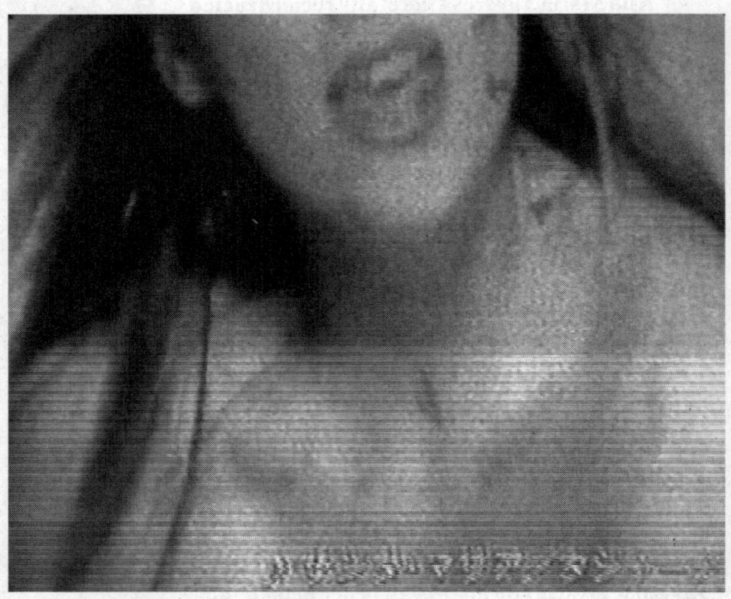

1.9   Geert Mul, *Tokyo FX*, 1995

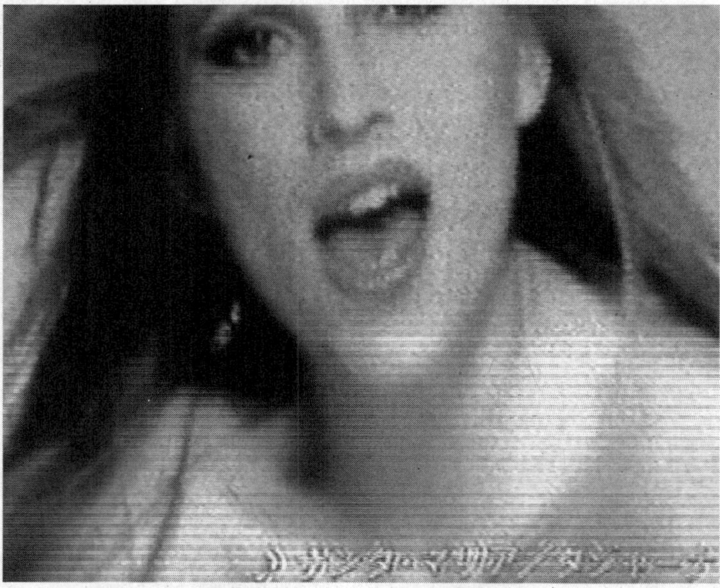

1.10 Geert Mul, *Tokyo FX*, 1995

repair the broken links between our consciousness and the world. He writes: 'not by striving for realism, which is already ubiquitous in our visual and information culture, but by reconnecting our numbed consciousness with reality' (see pages 59-84).

Ana Maria Tavares's carefully reconstructed non-places do precisely that: they bring our consciousness back into contact with reality. Anyone who has seen her work, like me, now has a different regard for the everyday reality of airports, waiting rooms and hotel lobbies. There are, of course, many more such examples. Heijne recognizes his ideas about the task of the artists in the work of an artist such as Germaine Kruip. In his essay 'Real Seeing' he quotes Kruip: 'You create reality as you look at it'. Heijne continues: 'If art can do anything – or to put it more emphatically, if art must do anything these days, it is that' (see pages 95-106).

Following Heijne's ideas about the task of the contemporary artist, one might imagine that he is dismissive of mass culture: but in fact the opposite is true. In a series of essays he has proved his pre-eminent ability to detect quality in popular culture as well as in the fine arts. The differences between art and mass culture are in any case much more complex than the polarity expressed in terms

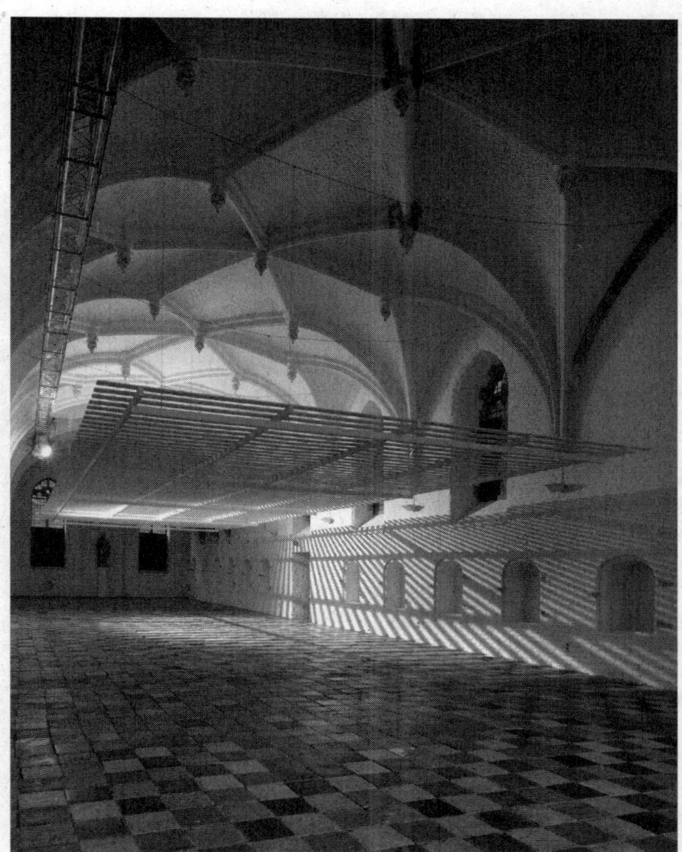

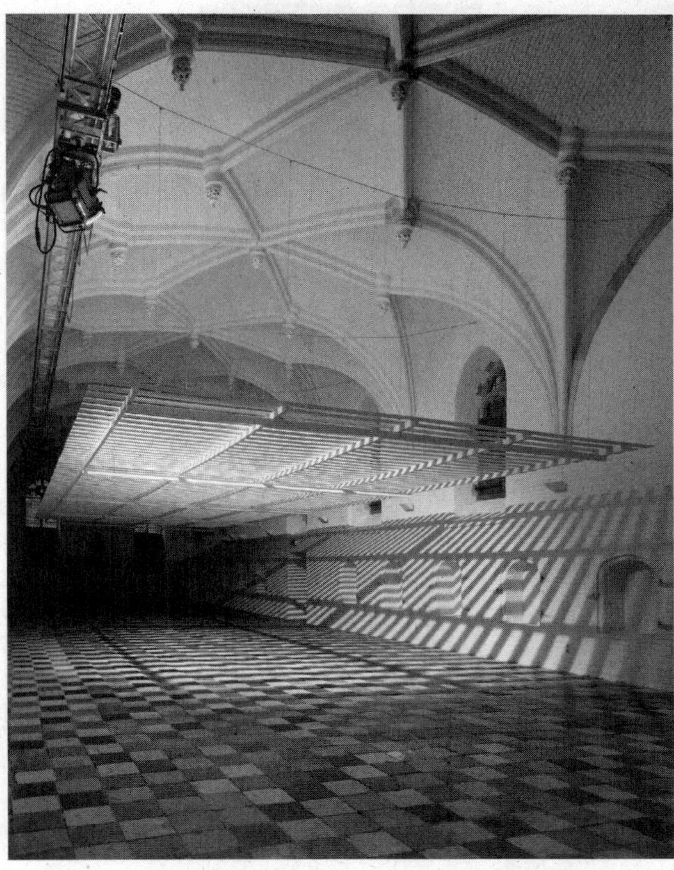

1.11 Germaine Kruip, *Counter Composition*, installation views, 2006

such as 'deep and meaningful' (art) and 'superficial and ephemeral' (mass culture).

Neither can one simply say that the hero of Coetzee's *Youth* represents significance and that the hero of Ellis's *Glamorama* stands for the loss thereof. It is true that the principal character in *Youth* is an insecure loser in search of profundity who has receded completely into his own world and that Victor Ward is a successful airhead entirely focused on the external world, but they also have similarities: Coetzee's romantic is every bit the narcissist that Ward is. He dreams not only of writing but also of the love and recognition that a poet's life will afford him. Rather than seeking a woman who loves him for what he is, he goes in search of a woman who loves the poet he hopes to become. Conversely you can also attribute romantic cravings to Ward: a longing for style and glamour, for the seductive world evoked by fashion, advertising, films and music videos, a hankering for yet another quick high, for life on the surface. In this respect Coetzee's romantic and Ellis's narcissist are not each other's opposites but rather two sides of the same coin.

Ellis's contemporary reality and Coetzee's 1960s reality also differ greatly but are not each other's opposites. *Glamorama* can obviously be interpreted as a critique of the loss of meaning

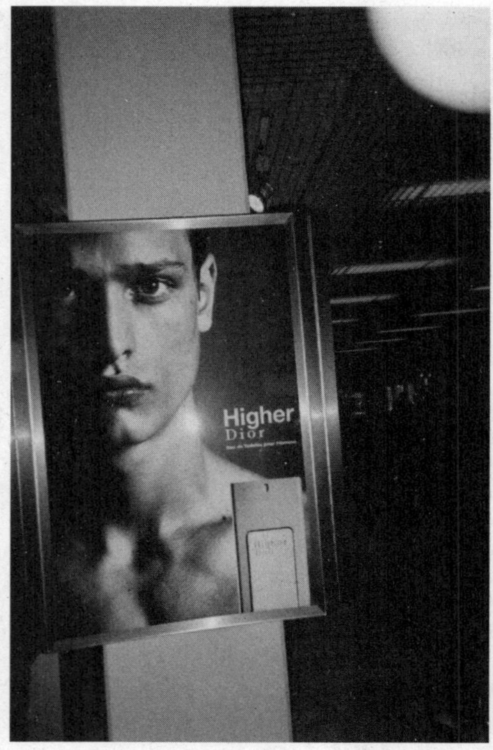

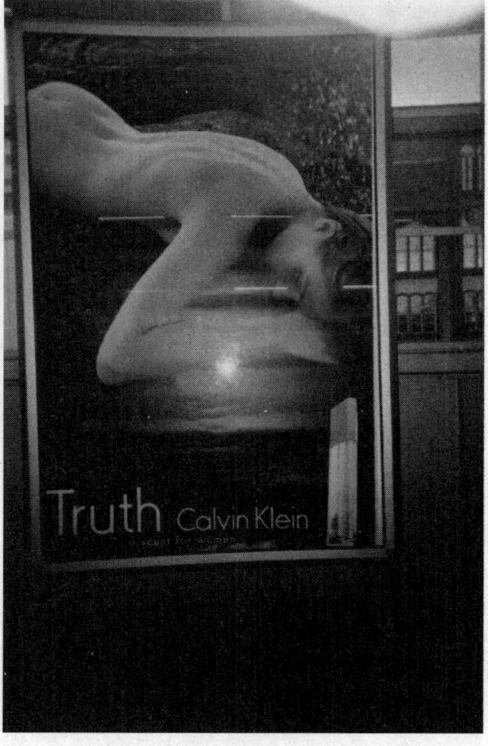

and authenticity in our modern reality; indeed *Glamorama* also has an irrefutably moralistic ending. But morality is not Ellis's forte. The power of his novel lies in his descriptions of the desire for a life of superficiality. It shows us that longing to lose oneself in a flow of seductive images is, in essence, deeply romantic. It is therefore facile to dismiss mass culture as synonymous with a lack of significance. Furthermore, Bas Heijne points out that beauty is no longer the preserve of art: 'popular culture is also aesthetic to its very core' (see pages 59-84).

If museums wish to discover how we can relate to modern reality they must take popular culture seriously. Simply reflecting upon mass culture from without – via art – is not enough. Museums must also research the influence that mass culture has from within. This immediately raises the question of how the museum – traditionally the temple of high art – can provide a platform for popular culture. This question is the central issue of the following section of this introduction.

2. Museums and Mass Culture

More and more contemporary artists are interested in the expressions of modern popular culture: fashion, advertising, music videos, new

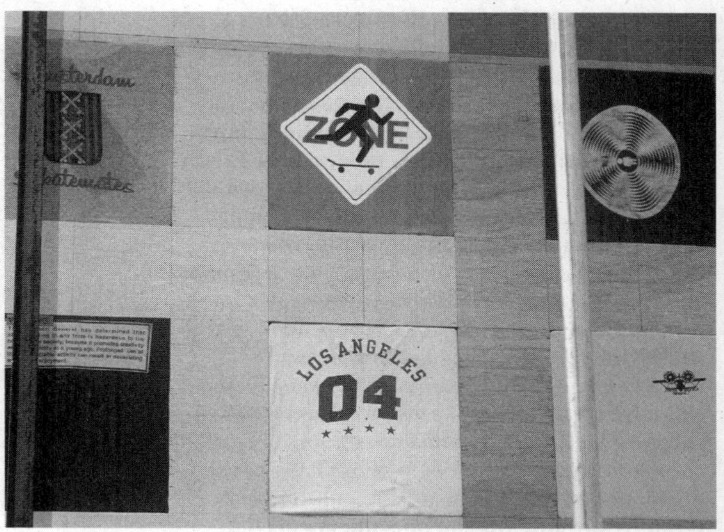

1.12 *This is the Flow*, Girl Skateboard Company, Gyz La Rivière & Robert Rosenau, 2001

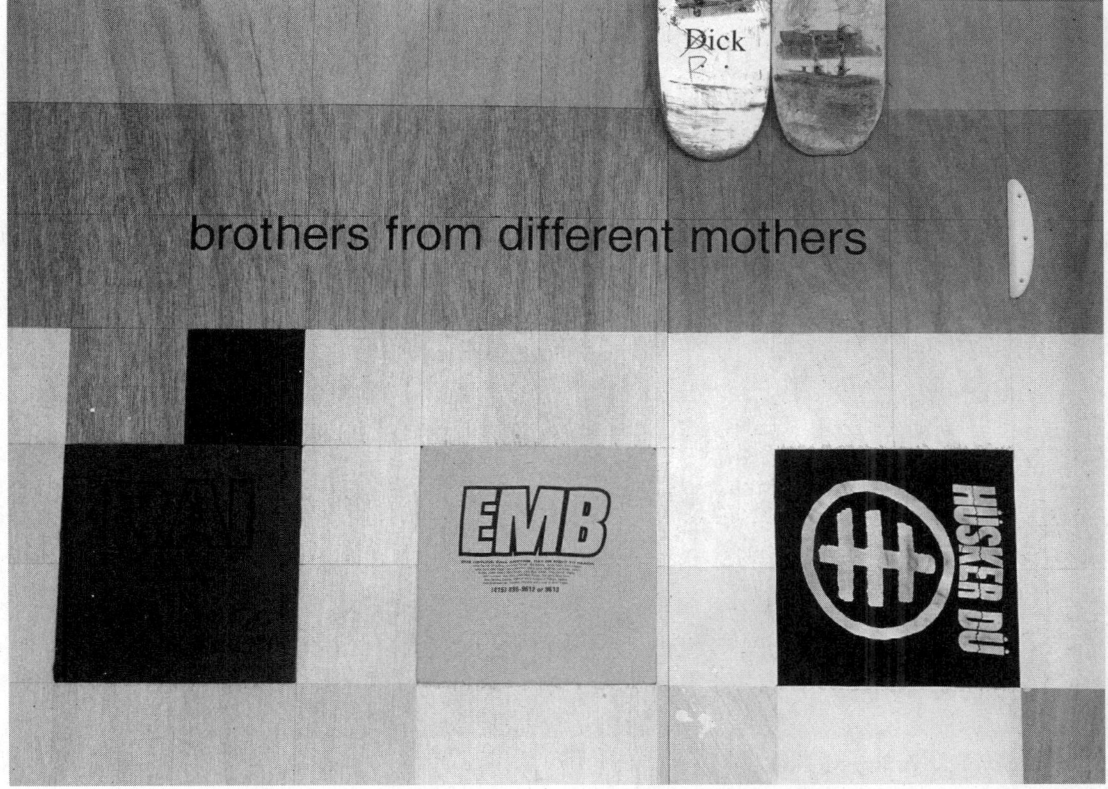

1.13 *This is the Flow*, Girl Skateboard Company, Gyz La Rivière & Robert Rosenau, 2001

media and design. But these disciplines also borrow increasingly from the arts. This continual cross-fertilization has blurred the boundaries between art and popular culture. Is this the future for art, or does it spell its demise? And how should museums relate to this problem? Should they stick to showing high art, or should they go in search of what art *might* become?

These questions were the focus of the collection of essays entitled *Kunst in crisis* ('Art in Crisis') published in 2003. In my eponymous essay, included in this volume, I outline the art world's reaction to the blurring of the boundaries between art and mass culture, which can be briefly summarized as follows: appropriation or negation. Some aspects of popular culture are proclaimed art and thus annexed by the art world. However, the majority of museums expressly ignore popular culture. This appropriation and dismissal is evidence of a deeply rooted cultural pessimism. 'Both strategies are designed to maintain, as much as possible, the distinction between high and low culture for fear of losing the universal values of high art. From this standpoint, art – and thus the museum as well – exists by virtue of the distinction between high and low culture, and the blurring of this distinction plunges art into a crisis' (see pages 115-130).

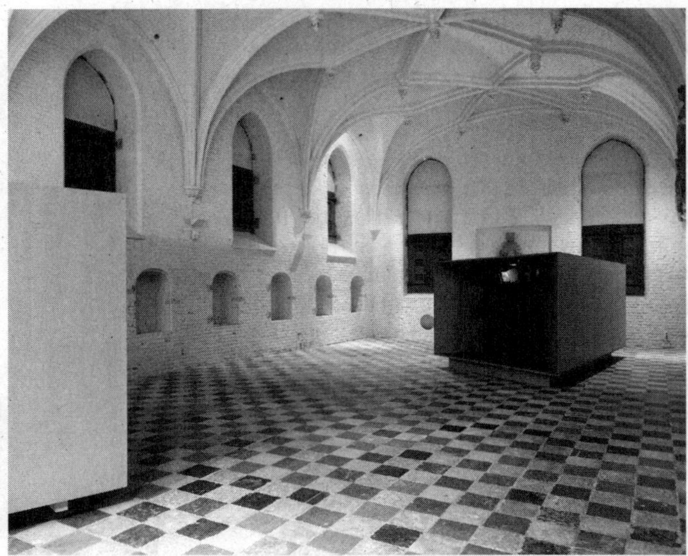

1.14 *Alexander Ranner & Cameron Rudd*, installation view, 2005

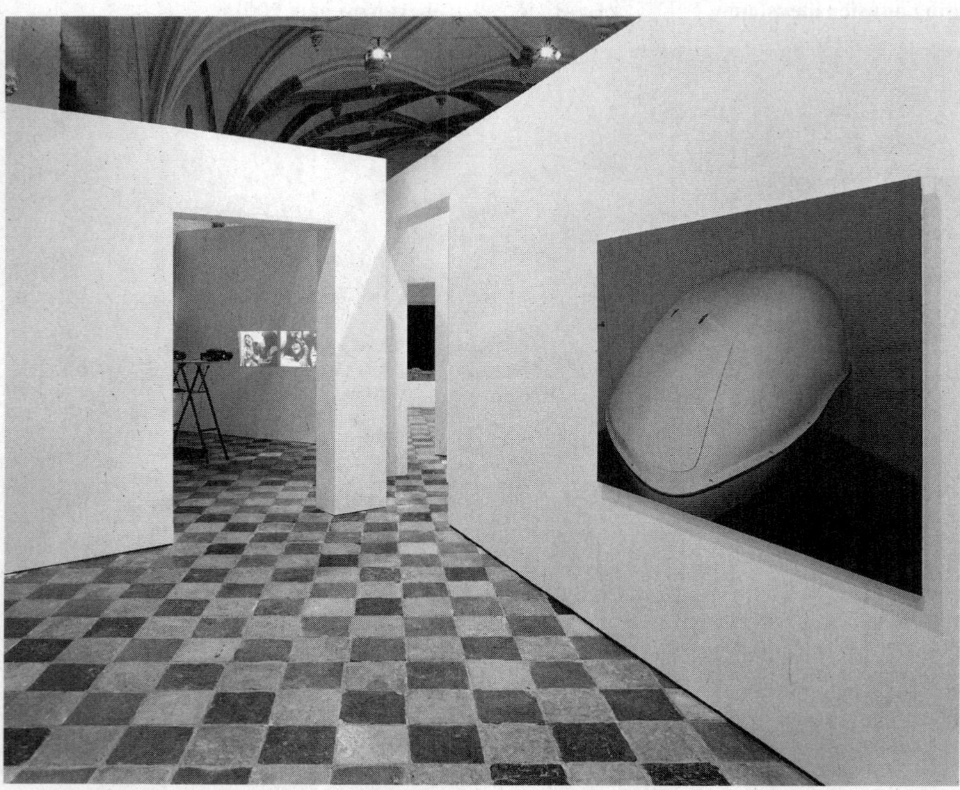

1.15 *De Werkelijkheid*, Janice McNab, 2004

Although the authors brought together in this collection had extremely divergent and occasionally opposing views, the Dutch art world's response to *Kunst in crisis* was unusually fierce. The book was seen as an attack on art; the title alone was viewed as yet another death knell for art. But the predominant interpretation was that art criticism – and not art itself – was in crisis. Museum directors, critics and curators no longer know how to relate to art. They employ populist tidings of doom about art for their own self-promotion. 'Art must simply be left in peace' was a common response. And, it was argued, popular culture has a place within museums simply by dint of artists' responses to it.

In spite of the strong reactions, the majority of the book's critics recognised that art has difficulty holding its own amidst mass culture. The fear that art as a component of classical culture has been demoted to merely a subculture for the (white) middle class is now widespread. Opinion varies only on what to do about this. The dominant view is that art must be protected from the influence of popular culture. For this reason, museums often distance themselves from mass culture and devote themselves exclusively to exhibiting high art. But this, I am afraid, makes all the more inevitable the doomsday scenario in which art's marginalization becomes complete.

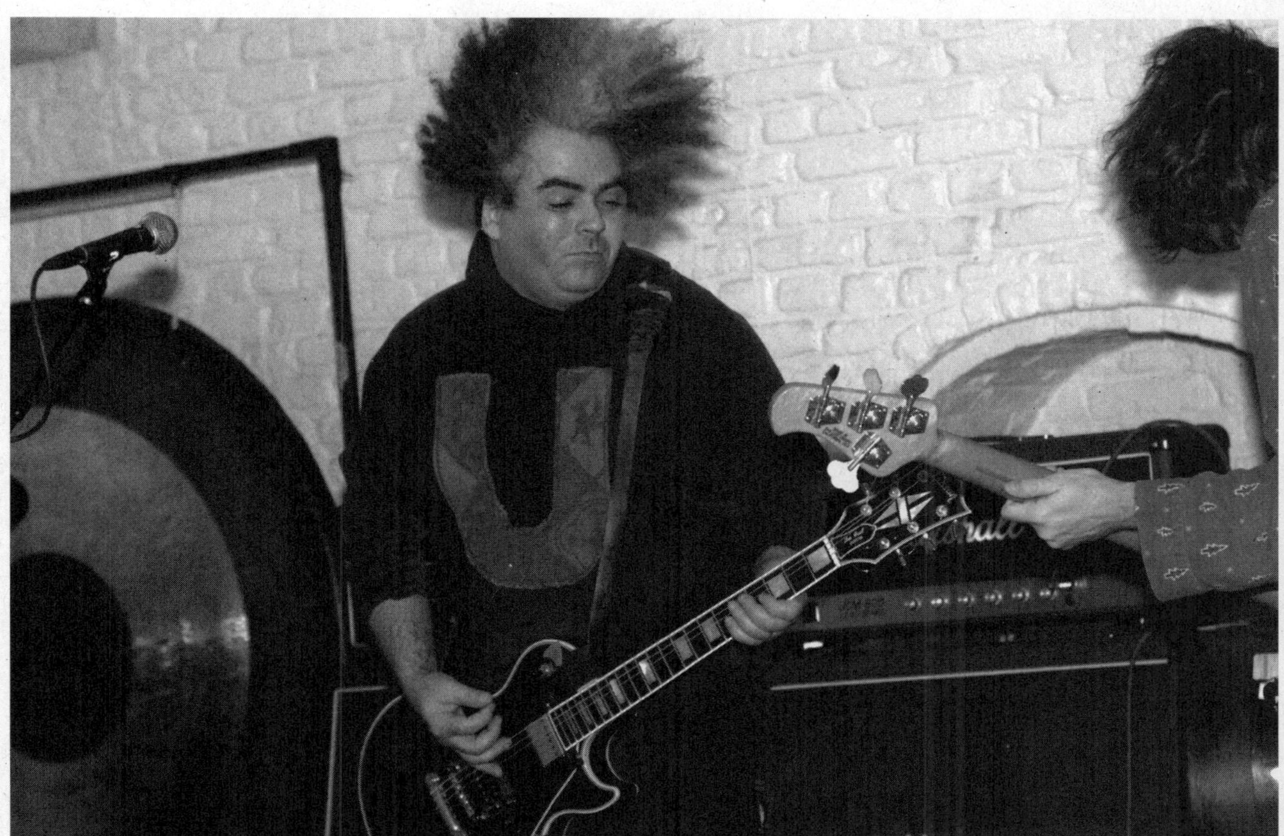

1.16 Live performance by The Melvins, De Vleeshal, November 12, 2004

Furthermore, museums are institutions uniquely equipped to give meaning to developments within popular culture. But how can museums make room for 'low' culture without developing an identity crisis? And in doing so, how can we assure art's autonomy? Towards the end of my essay 'Art in Crisis', I provide the beginnings of an answer to these questions. In order to redefine the museum we must unfetter ourselves as much as possible from the idea that the art history perspective is the only framework within which cultural products can be displayed.

The art history perspective has become exceedingly insular. The majority of themes for exhibitions are borrowed from art history, and works of art are displayed in the context of styles and movements in art history. There is nothing wrong with this. The appreciation of a work of art can benefit greatly from the art history perspective, but usually this is stressed to such an extent that it seems as though art is (or can be) only about art. The fact that the art history perspective has led to a form of tunnel vision is also clear from many museums' neglectful relationship to technological developments. The emergence of countless distinct institutions for so-called 'new media art' is a symptom of this malaise. These separate venues would have been unnecessary had the art history perspective

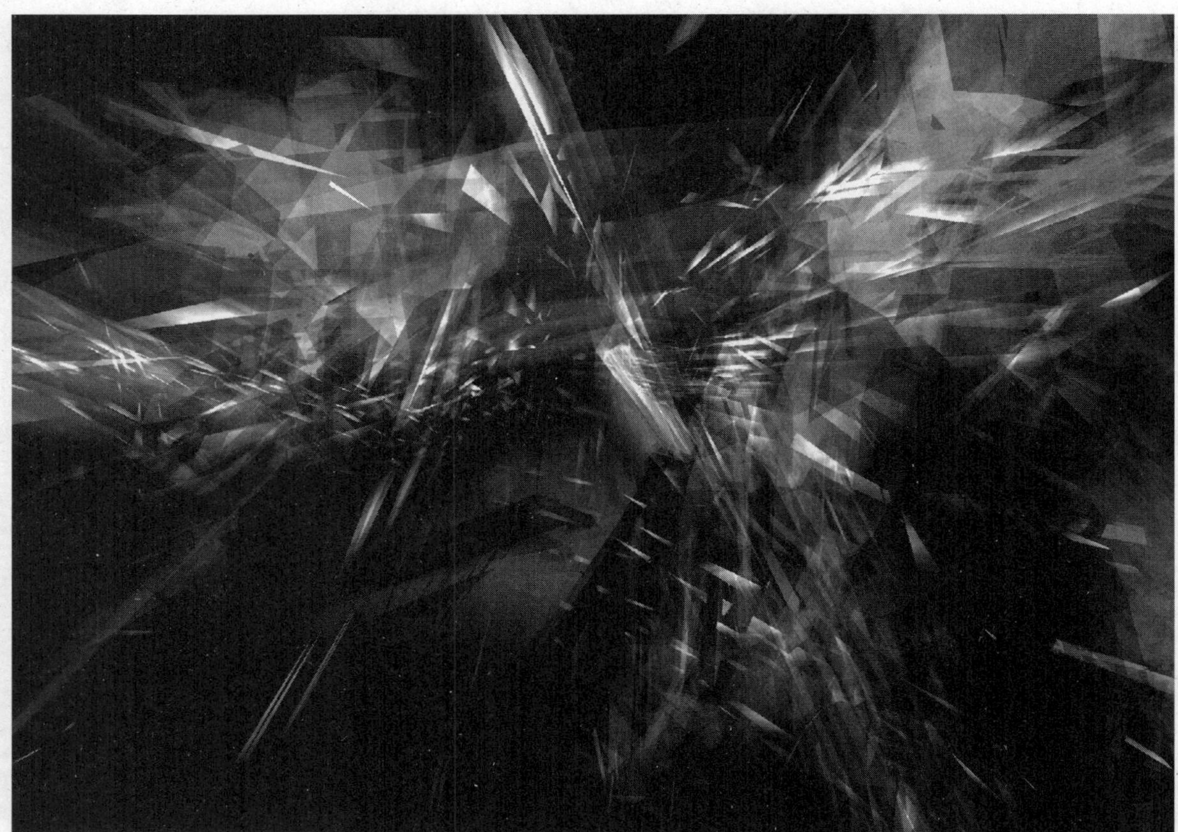

1.17 Lucas van der Velden, *Meta_Epics*, 2007

granted technology more space within the museum. 'The intrinsic position of technology has been banished from art history', comments Geert Mul on this subject in an interview (see pages 131-143). He regrets this development, because, he says, 'it's important that a museum of modern art reflects the current language, and that's something modern art museums definitely do not do. They prefer to reflect upon a high point of painting from the Golden Age, whereas at that time they *did* paint and work in the language of the people. The greengrocer's sign was not too far removed from the painting by the artist, who used the same vocabulary and the same medium, but in a different way.'

Being an artist is above all a question of mindset. However, because of art history's tunnel vision, this mindset is often not recognized initially. Lucas van der Velden is an example of someone who operates largely outside the art world. Among many other projects, he has developed software that enables computers to generate sounds and images. 'Artists who work with computers seem to be more involved with the medium itself than other artists,' he says. 'Perhaps this is because the computer has only been in use for a quarter of a century. It all seems new, and it appears to be more difficult than painting, but if you look at painters who've pored

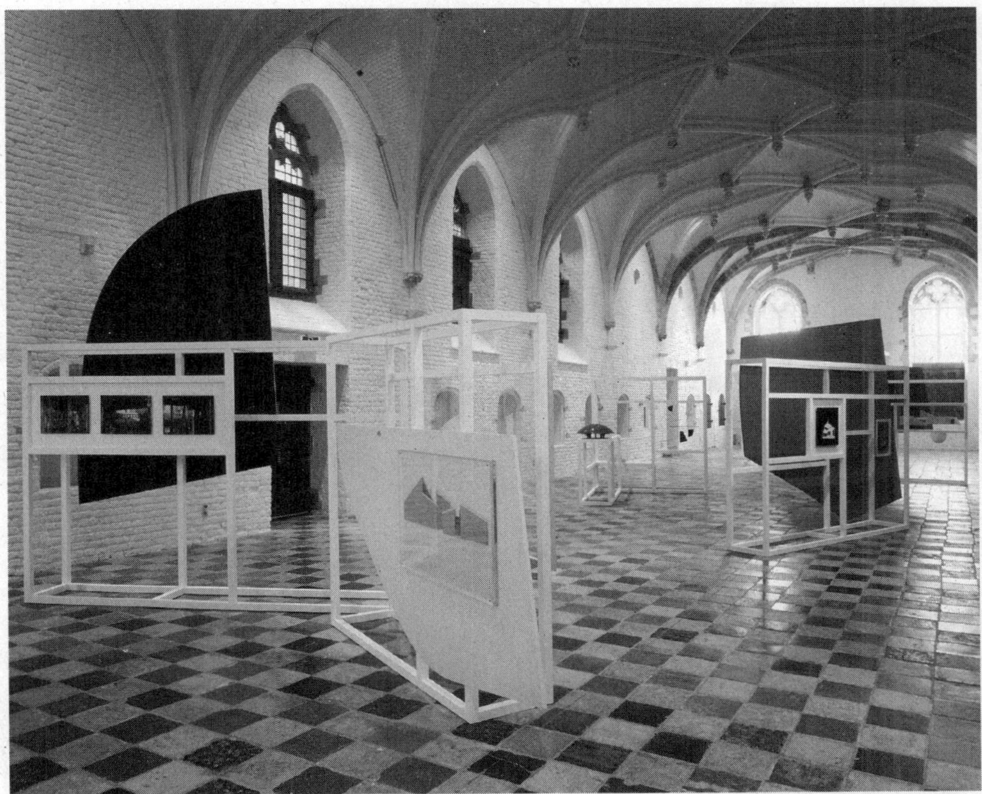

1.18 Toby Paterson, *Broken Arabesque*, installation view, 2006

over their brushes and paint mixing for years on end, you see the same devotion – for that is what it is. It's not that the technology is more important than the outcome. On the contrary, I think that understanding the means by which you work leads to better work. The medium you work with has to have some intrinsic connection to the final product.'[4]

Seen from the perspective of art history, the museum is for art, and vice versa. There is no place for non-artists. But with the blurring of the boundaries between high and low culture, the distinction between who is and who is not an artist is less easily made and is indeed hardly relevant. I deal with this matter at length in my essay 'Art in Crisis'. Chris Darke also refers to this in his article about the video artist Chris Cunningham, in relation to the presentation of his first autonomous art work, *flex*. When this work was first exhibited, 'much was made of Cunningham's "commercial" background, his evident ease and facility with special effects, as though the moving image in a gallery still needed to be low-tech and low-res for the sake of "authenticity"'.[5] But 'what about other artists such as Matthew Barney, Doug Aitken (himself a former maker of music videos), Mark Lewis, Isaac Julien and Eije-Liise Ahtilla? All are major art-world names, and all pursue a degree of technical "finish" that makes no effort

[4] Fragment from an interview that Valentijn Byvanck and I conducted with Lucas van der Velden in 2001 in relation to his exhibition *0010* at De Kabinetten van de Vleeshal, March-May 2001. The exhibition included music videos he made for the bands Endorphins and Eaven (under the name 0030 in collaboration with Martijn van Boven), Speedy J, Eog and Phako.
[5] The work was first exhibited as part of the group exhibition *Apocalypse: Beauty and Horror in Contemporary Art* at the Royal Academy of Arts in London in 2000. It was presented at De Vleeshal in March-April 2001.

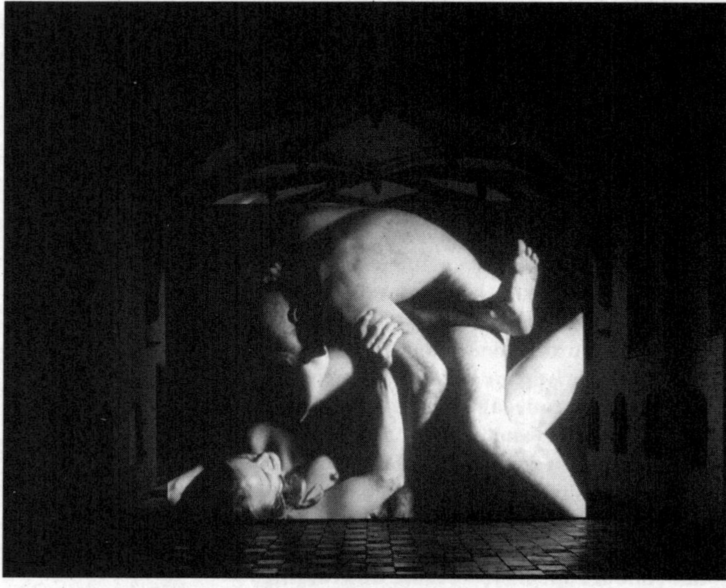

1.19 Chris Cunningham, *flex*, installation view, 2001

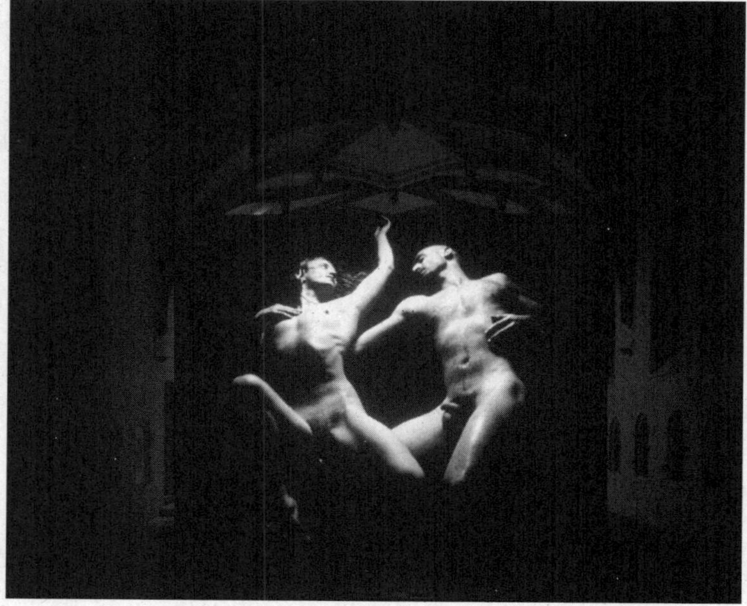

1.20 Chris Cunningham, *flex*, installation view, 2001

to conceal the rigorous professionalism of their work' (see pages 183-192).

Once museums discard their art history perspective, the (impossible) question as to who is an artist and who is not becomes as unimportant as the untenable distinction between low and high culture. Museums are no longer restricted to exhibiting only the work of visual artists. As long as the most important condition for making art is guaranteed – the independence of the maker – this poses no threat for artists, for art or for the museum. This is possible by linking the independence of art not to the person who makes it – the visual artist – but instead to the site in which it is exhibited, i.e. the museum. In this way non-artists can also make use of this autonomy. Within this vision the museum is not principally a place for art and artists, but rather a space for ideas. In the museum as a 'space for ideas' the tunnel vision of the art history perspective can be avoided, and room can also be created for interesting developments that take place beyond this perspective. In this way the museum can truly highlight and give meaning to developments within our modern mass culture.

The project *Higher Truth No. 5*, about fashion's power to seduce, is a good illustration of this. Each fashion label suggests its own parallel

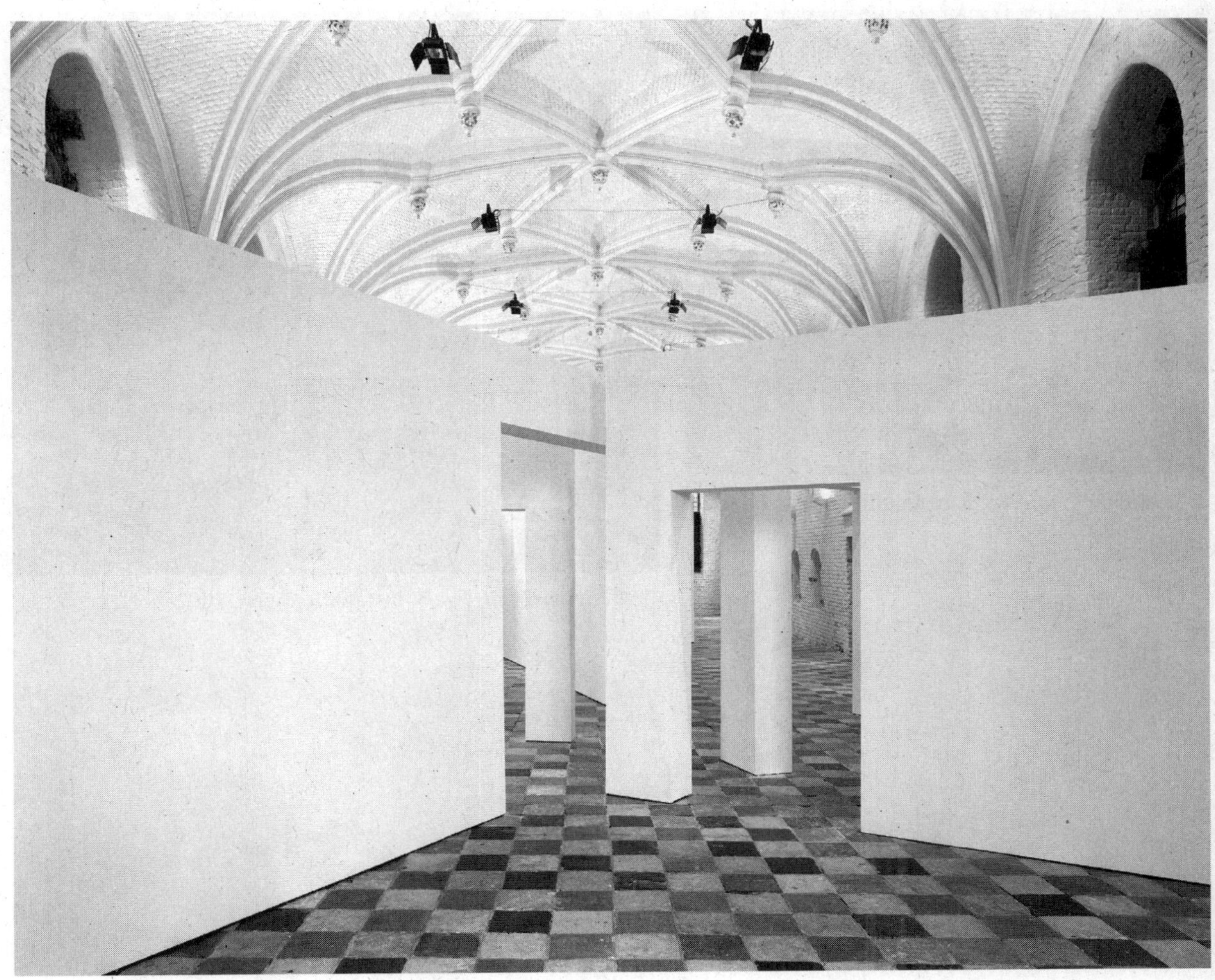

1.21 *De Werkelijkheid*, installation view, 2004

universe. Adverts, boutiques, shows and models – each of these serves the creation of a purely visually defined, more beautiful, more stylish and better world. The imagery employed to evoke this world is perhaps fashion's most fascinating aspect. But this is barely touched upon when fashion is exhibited in the museum. The fact that museums always present fashion in the form of garments, and preferably within a craft context, is designed to legitimize the place of fleeting and ephemeral fashion within the temple of art. But this point of view completely ignores precisely that which makes fashion so interesting.

The exhibition *Higher Truth No. 5* attempted to show the seductive power of fashion in its purest form. The exhibition consisted of a 1:1 scale model of a luxury fashion boutique, but without the clothes. Visitors were received by hand-picked, carefully trained, extremely attractive and specially dressed shop assistants. The design of the boutique, the special lighting and the staff conjured up a unique reality in spite of the absence of a physical product. *Higher Truth No. 5* set out to demonstrate fashion's ability to generate a convincing ambience that can be neither rationally described nor explained. Fashion, in other words, as sensation, seduction and intoxication. In her article about this exhibition, Anna Tilroe writes, 'we see an exhibition that is not an exhibition,

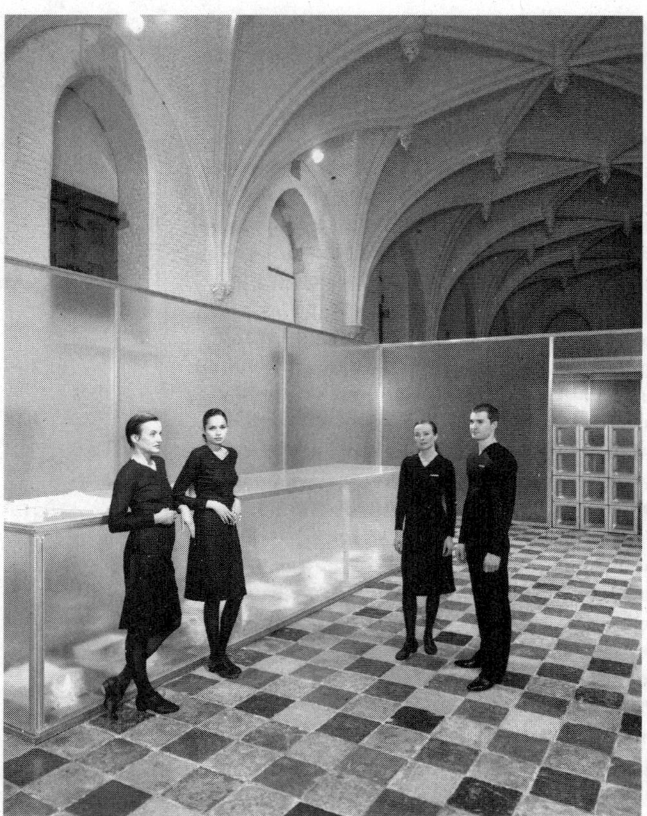

1.22 *Higher Truth No. 5,* installation view, 2002

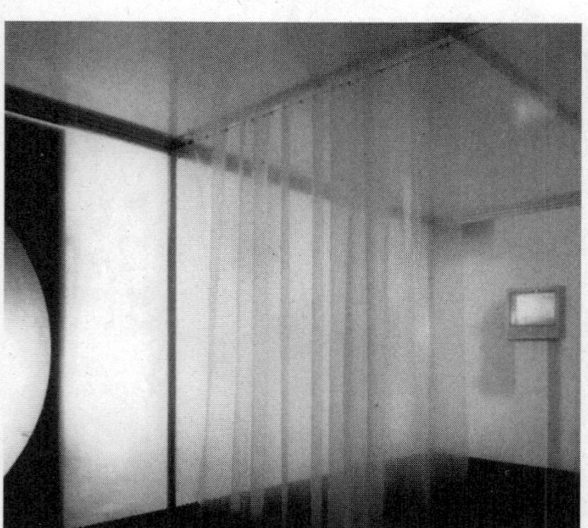

1.23 *Higher Truth No. 5,* installation view, 2002

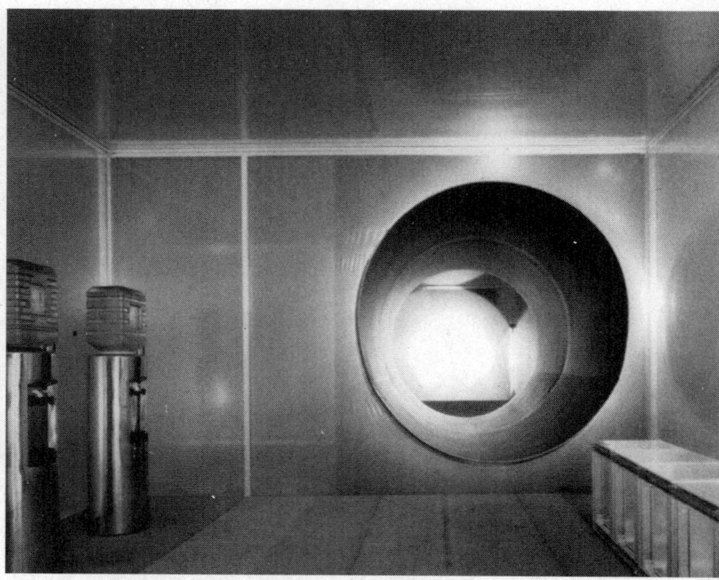

1.24 *Higher Truth No. 5,* installation view, 2002

but rather a magnified reflection of the excessive way in which we have lost touch with reality.' (see pages 145-152).

Higher Truth No. 5 is relevant here because it was not a project initiated by artists but by people from the world of fashion. The project involved a team that included Guus Beumer and Alexander van Slobbe, responsible for the fashion labels SO by Alexander van Slobbe and Orson & Bodil, and Herman Verkerk and Rianne Makkink, both architects with experience in fitting out fashion boutiques and designing fashion shows. Of course there are also artists who contribute germane comments about fashion, but in this case I found the vision of people from the fashion world itself somewhat more interesting. They are, after all, the ones who give form to fashion's seduction. Beumer and Van Slobbe have never shied away from commenting upon fashion itself in their collections and shows. For example, they responded to the lack of innovation in the clothing industry by presenting a collection as a purely schematic vision with no finished product. And their next collection played with the idea of topicality by assigning fictional dates to works from earlier collections. But opportunities for such critical commentary are restricted by their reliance upon the market. The museum offered them far greater autonomy with which to reflect upon the world of fashion.

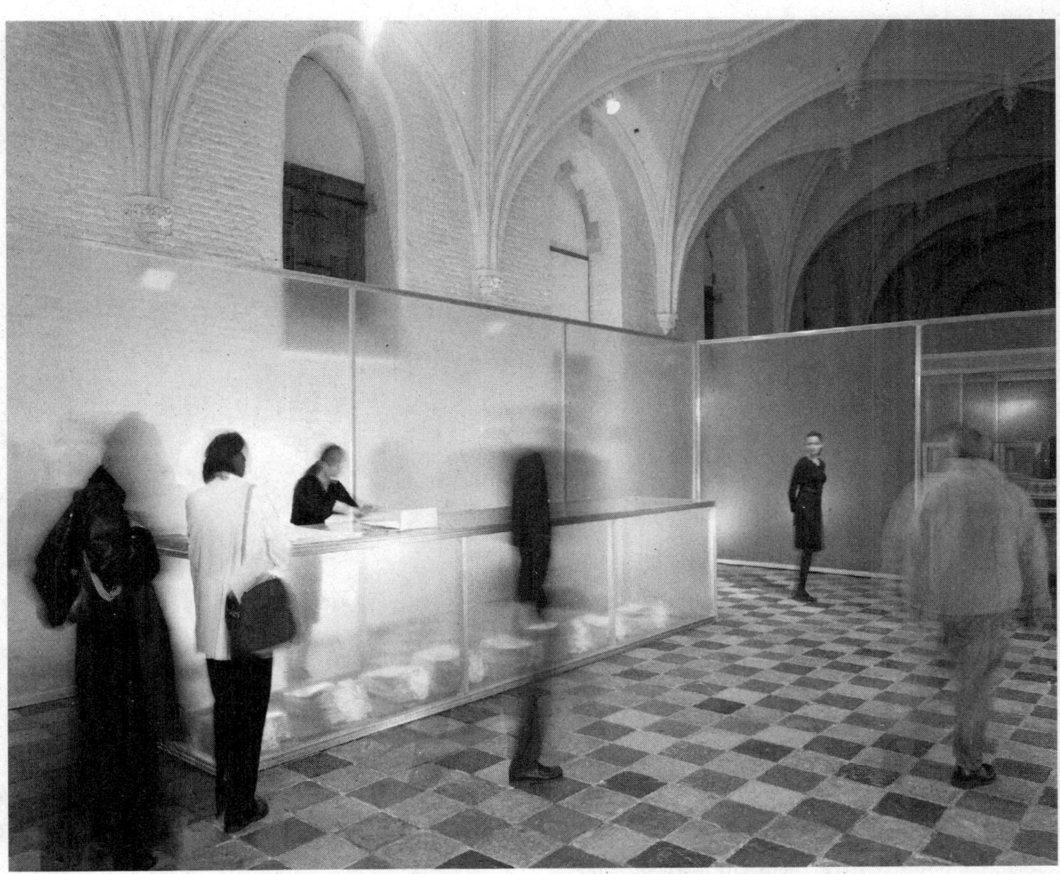

1.25 *Higher Truth No. 5,* installation view, 2002

Fashion is a fascinating and topical subject, not only in its own right but also within a broader context. Guus Beumer demonstrates how fashion's seductive strategies are now being applied in other areas. In his opinion the language of fashion has moved far beyond its own boundaries and has colonized the world of architecture (see pages 153-182).

Another example from this same series of projects is *Generating Live* by Geert Mul and others. At the end of the 1990s Mul began a series of visual experiments in nightclubs. Because there was no name for what he did, he described it as 'a show with TV screens: light and pictures at the same time'. Only later was this dubbed VJ-ing: projecting images to accompany DJs in clubs. The work of some VJs is indebted to the visual arts and many of them, including Geert Mul, have studied at art schools. An interview with Mul included in this volume makes apparent that the clubbing context sets limits to VJs' artistic freedom: eventually it comes down to bar profits. A museum can offer much greater autonomy. For this reason I found it interesting to ask Geert Mul to translate his experience of working with sound and images in clubs to the museum context of De Vleeshal, where there is also space for discussion and reflection. Or as Geert Mul puts it, 'the reason I have always had and shall always retain a soft

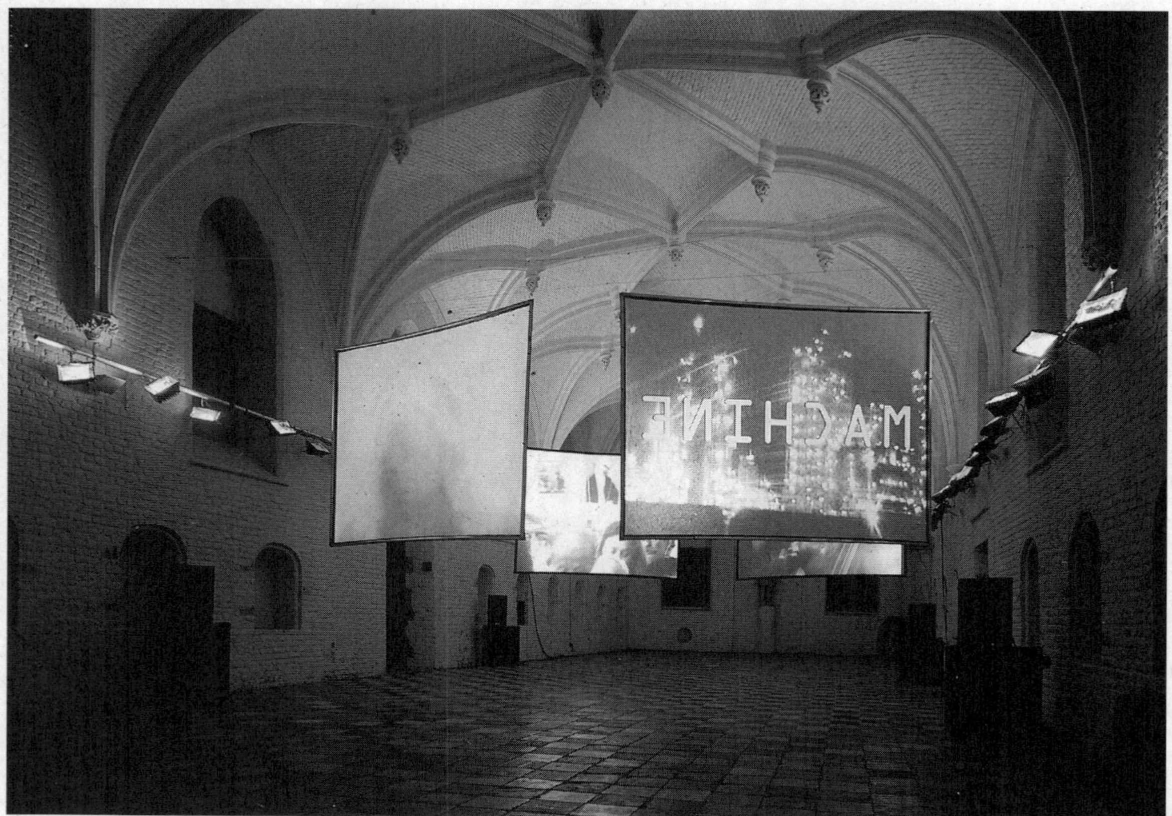

1.26 Geert Mul, *Generating Live*, installation view, 2000

spot for the official art circuit is the level of the discussion conducted there. The dialogue, the discussion, is after all what it is all about. That is an important distinction to pop, which is always about the icon and the individual' (see pages 131-143).

Because artists almost always have a free hand, museums are more tolerant than other arenas. In his essay Chris Darke points out that the music video director Chris Cunningham was able to realize his first independent art work, *flex*, only for a museum, albeit with funding from a commercial gallery. Cunningham had originally begun the work as a commission from the British television broadcaster Channel 4, but the content was so explicit that they backed out (see pages 183-192). It also seems that some artists interested in disciplines related to visual art have nonetheless consciously decided to work within the visual arts because the museum context offers them greater freedom. For example the work of Krijn de Koning, Hendrik-Jan Hunneman, Toby Paterson and Tomas Saraceno occupies the border zone between art and architecture. By satisfying their love for architecture through visual art they avoid the more banal aspects of architectural practice, such as dealing with clients and working within building regulations.

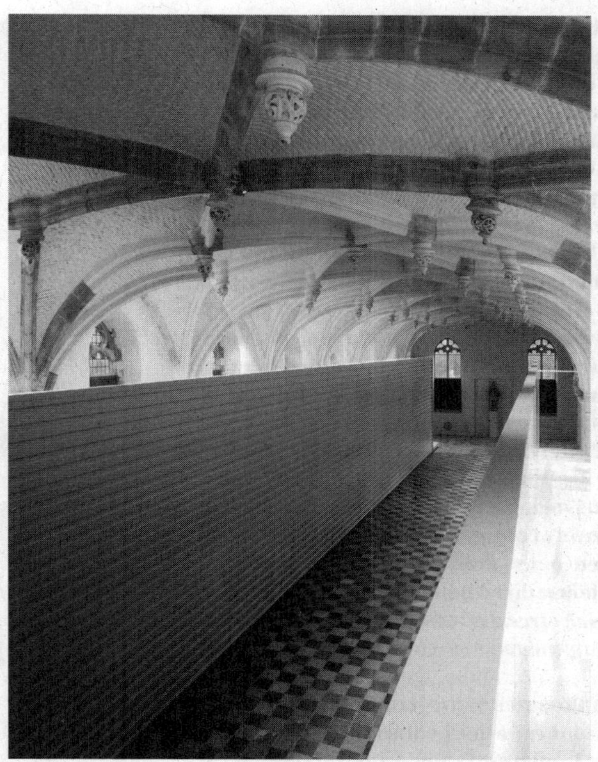

1.27 Hendrik-Jan Hunneman, *Kiss & Go*, 2002

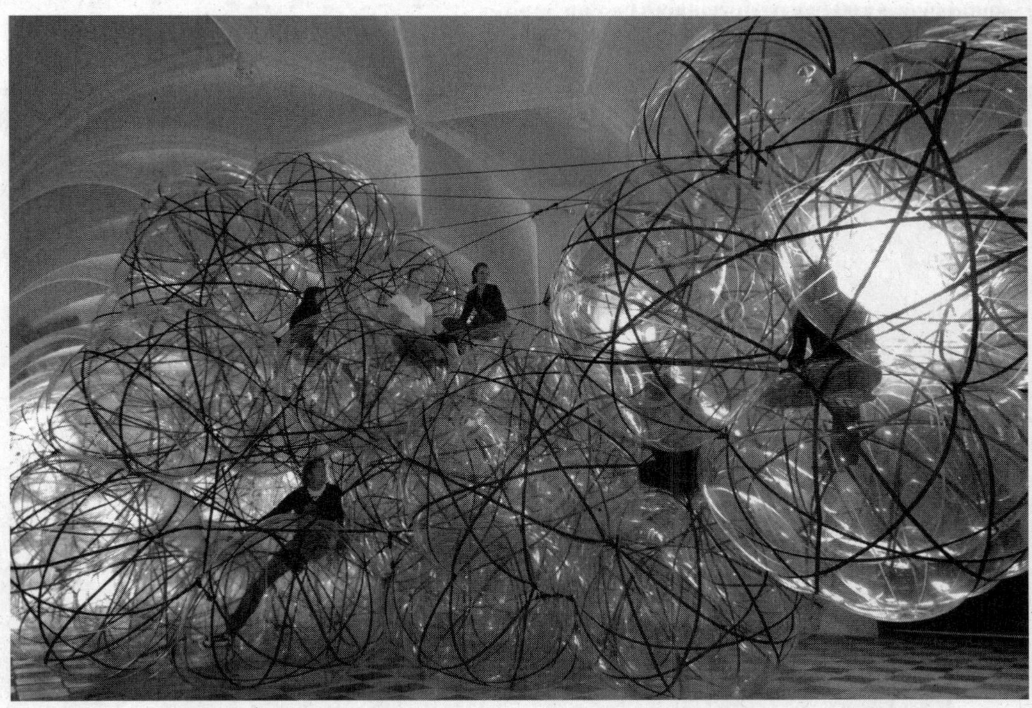

1.28 Tomas Saraceno, *Air-Port City*, 2007

In fact all disciplines related to the visual arts place certain restrictions on their practitioners. For this reason, it can also be interesting for choreographers and musicians to develop their projects within the liberal context the museum can offer them. Although they may be able to work with a relative degree of freedom in their own professional field, the museum offers the opportunity to stage a performance without a beginning and an end, and to forgo the need to release their work in CD format. It was within this framework that the choreographer Krisztina de Châtel realized an extraordinary dance project in collaboration with the visual artist Birthe Leemeijer in De Vleeshal. *Entree* was a piece of choreography for the audience: each visitor was given the opportunity to experience De Vleeshal's imposing Gothic space as a dancer. During a period of three months a team of dancers continually performed the piece, and members of the public were encouraged to join in and received help in picking up the movements. *Entree* dispensed with the distinction between dancers and audience.

The same series of projects included screenings of three films by the multi-faceted visual artist Cameron Jamie, with live performances of the soundtracks by The Melvins and sound pieces and live performances by Pan Sonic. Jamie's films deal with social rituals – or what he

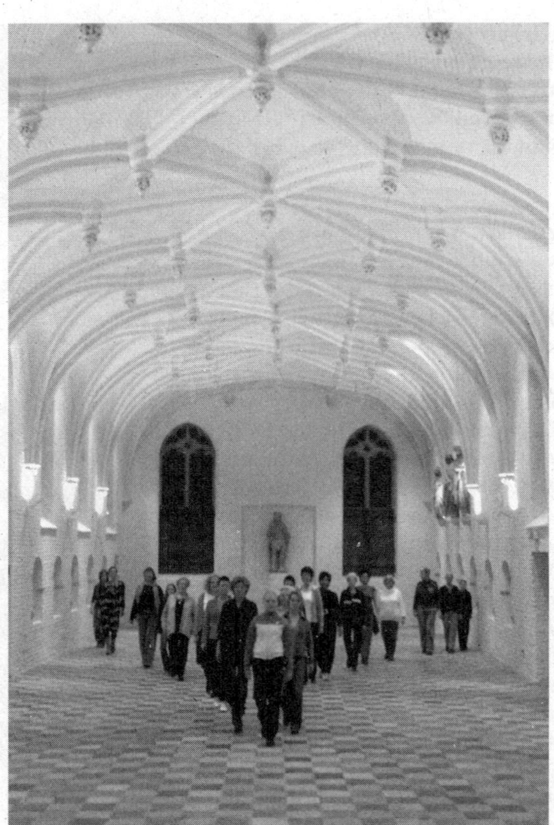

1.29 *Entree*, installation view, 2006

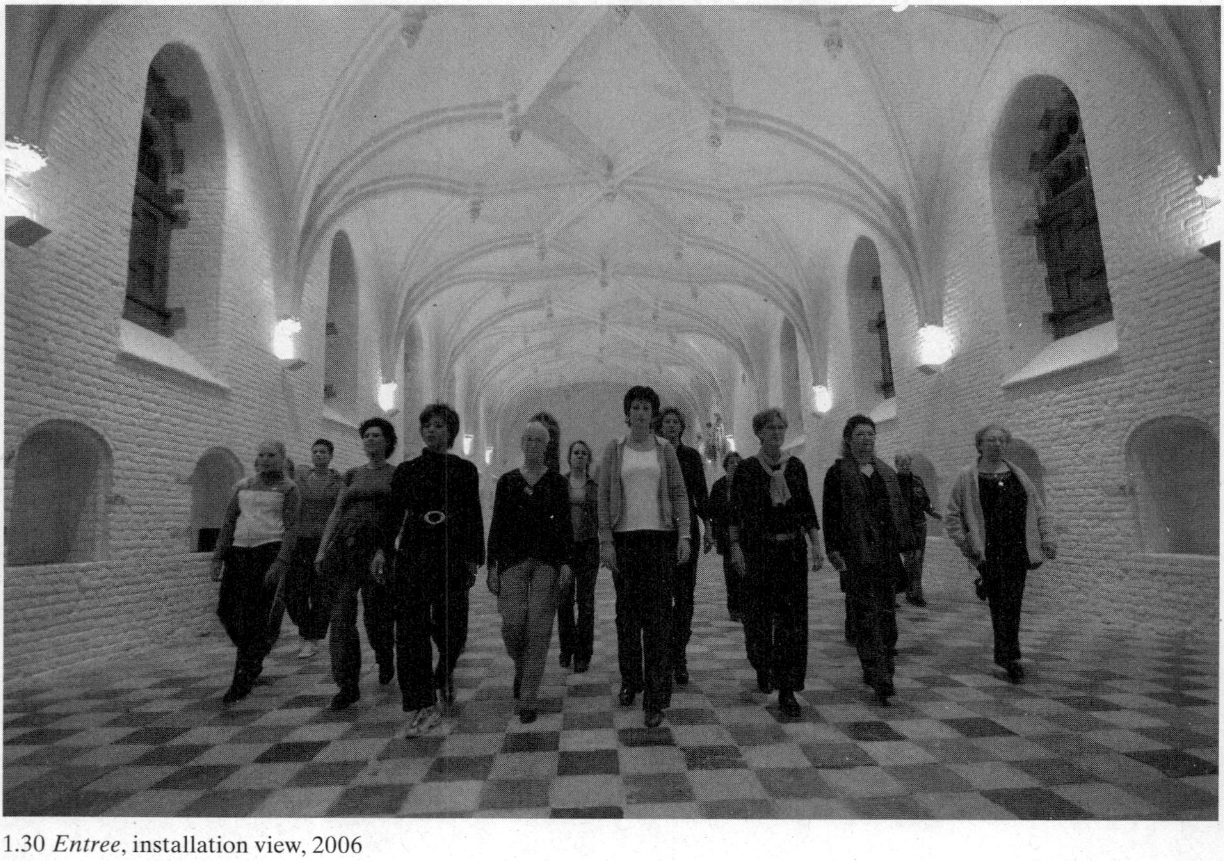

1.30 *Entree*, installation view, 2006

himself calls 'social theatre'. He is interested both
in age-old rituals such as the Krampus in Austria
and in new subcultural rites such as backyard
wrestling. In *BB* Jamie documented teenagers in
Los Angeles who re-enact televised pro-wrestling
matches. However, these bouts lack the gloss and
regulations of professional wrestling and involve
teenagers diving off garage roofs and attacking
one another with ladders and garden furniture.
In combination with the powerful, claustrophobic
soundtrack by The Melvins, *BB* is a penetrating
representation of a phenomenon that Jamie
describes as 'unsettling'. The Melvins' usual
performance circuit does not naturally lend itself
to such a collaborative venture. Museums are
much more open to such a project; indeed almost
every time this project has been performed it has
been at the behest of a museum.

The instrumental music of Pan Sonic (Mika
Vainio and Ilpo Väisänen) is characterized by
its impressive atmospheric scope. Their often
minimal compositions, principally produced on
analogue sound generators – some of which they
built themselves – vary from spell-binding ambient
pieces to merciless and almost unbearable sonic
assaults. Their sound piece was inspired by De
Vleeshal's former function as a meat market.
Large meat hooks dangled at 50-cm intervals on
steel wires from a 26-metre steel bar. The hooks,

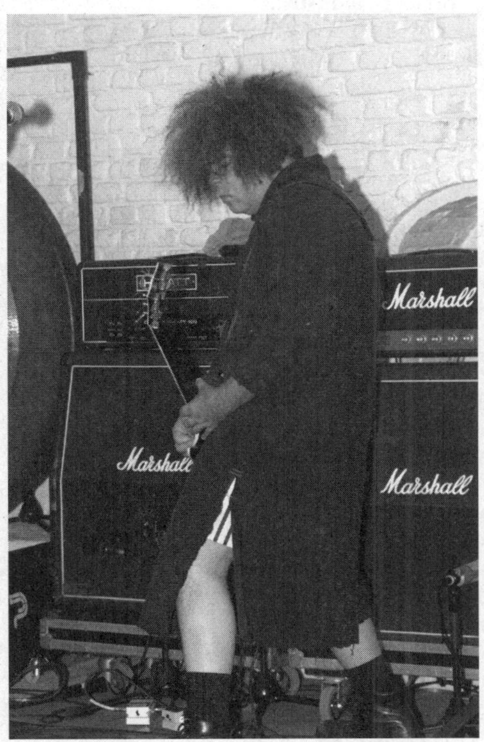

1.31 Live performance by The Melvins,
De Vleeshal, November 12, 2004

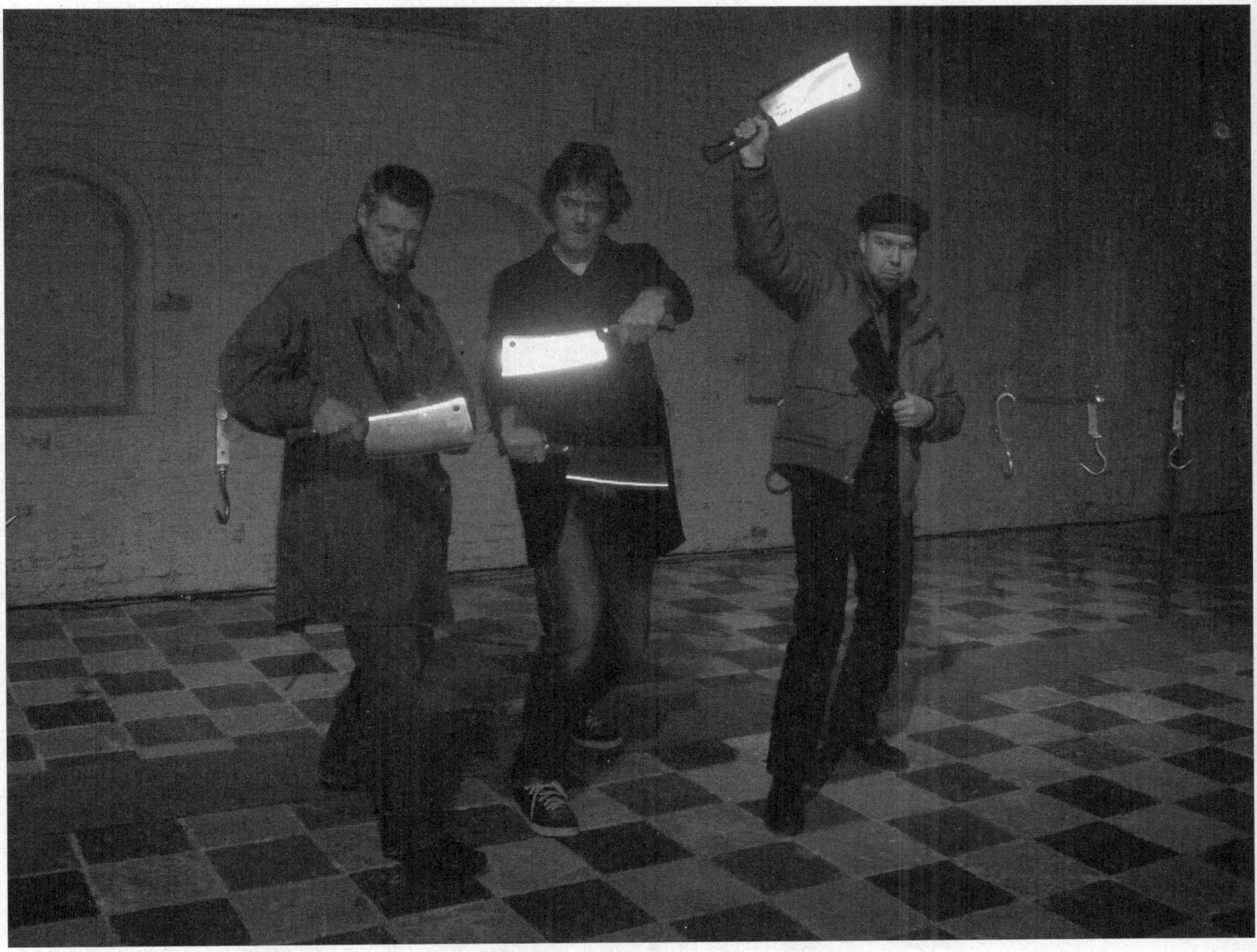

1.32 *Pan Sonic*, Ilpo Väisänen, Rutger Wolfson and Mika Vainio, De Vleeshal, 2005

which hovered just above the floor, were blown against one another by two large industrial ventilators placed at either end of the space. The sound made by the colliding hooks was manipulated to create an insistent acoustic soundscape. For the opening, Pan Sonic gave a one-off improvised performance using the sounds generated by the installation.

Perhaps the most far-reaching project in this series, in that it presented street culture within the museum context, was the exhibition *This is the Flow*, on the creativity of skateboard culture. There are interesting parallels between the art world and the skateboarding scene. In the latter, integrity is perhaps an even more important concept than it is in the art world. This integrity is not connected to an anti-commercial stance but rather to the possession of a personal style and/or creativity. Many talented Americans have found the first outlet for their creative ambitions around skateboarding, for example through producing skateboard films, photographs and fanzines. Spike Jonze, the director of films such as *Being John Malkovich* and *Adaptation*, is one of the most famous examples of this. Jonze is also a co-founder and co-owner of the Girl Skateboard Company. The aim of *This is the Flow* was to provide an insider's perspective on the creativity of skateboard culture. De Vleeshal invited the

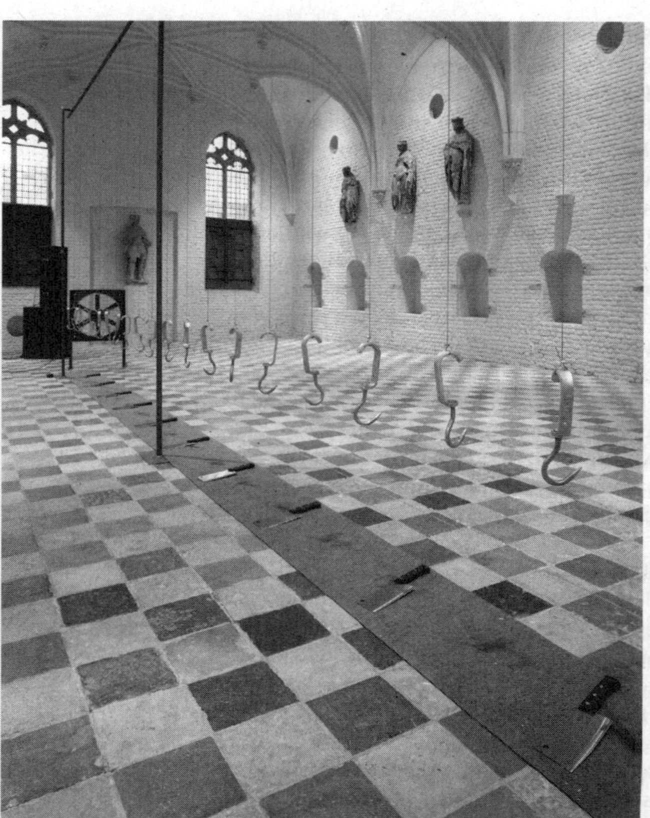

1.33 *Pan Sonic*, Ilpo Väisänen and Mika Vainio, 2005

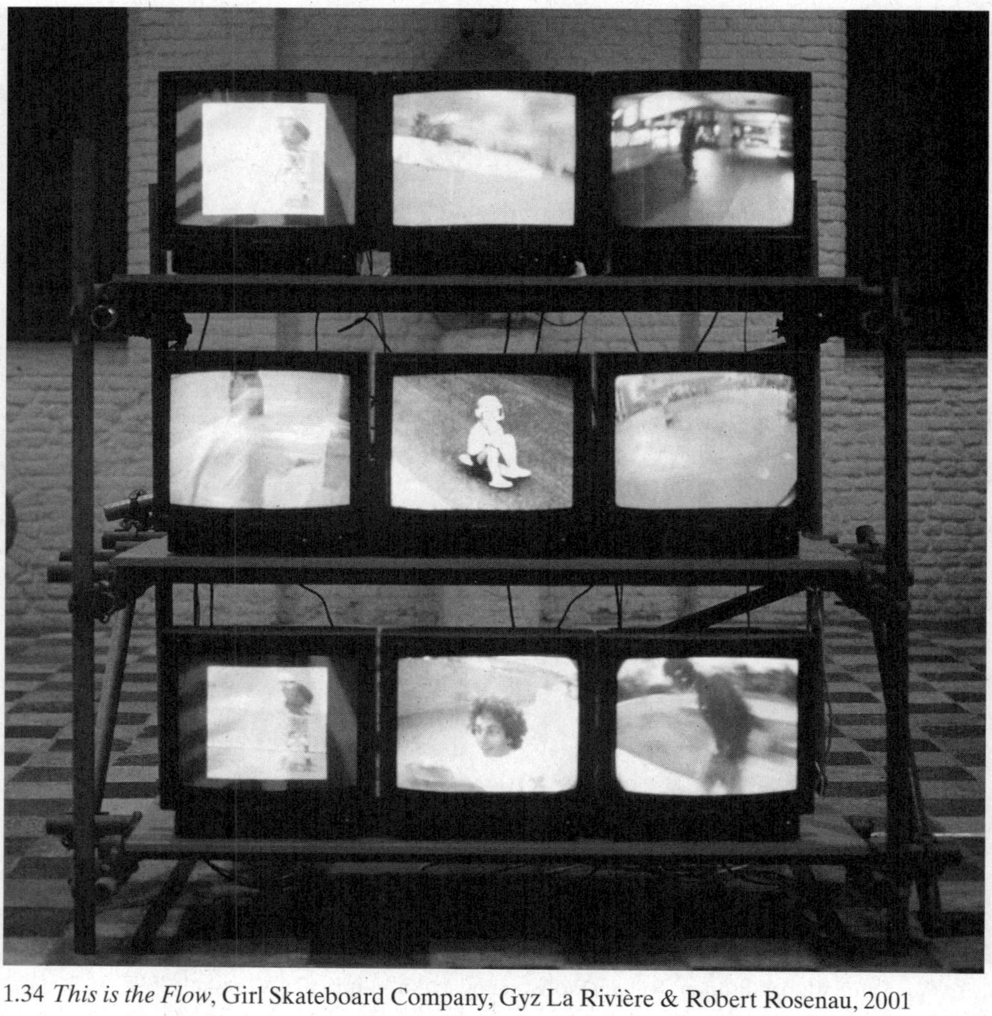

1.34 *This is the Flow*, Girl Skateboard Company, Gyz La Rivière & Robert Rosenau, 2001

Girl Skateboard Company's design team, known as the Art Dump, and artists and skateboarders Gyz La Rivière and Robert Rosenau to make an exhibition. La Rivière and Rosenau created installations in which they used their collection of skateboard paraphernalia to visualise their obsession with skateboarding. The Art Dump designed a series of 10 new skateboards especially for the exhibition and provided an insight into the creative process that led to their creation. This *De Vleeshal series* was put into production and distributed worldwide.

The definition of the museum as a 'space for ideas' offers museums not only the room to give meaning to developments in mass culture from within, but also the opportunity to reflect upon developments taking place in society.

3. Museums and Their Role in Society

In his visionary novel *Cocaine Nights*, J.G. Ballard paints a disturbing picture of the near future.[6] In a hypermodern holiday resort on the Costa del Sol, total apathy reigns. Luxury white air-conditioned holiday villas stand baking in the sun. Everyone sits indoors with the curtains closed, still groggy from the sleeping pills they took the night before. Nobody participates in any of the many recreational activities on offer. A new tennis

[6] J.G. Ballard, *Cocaine Nights*, Harper Collins, London, 1996.

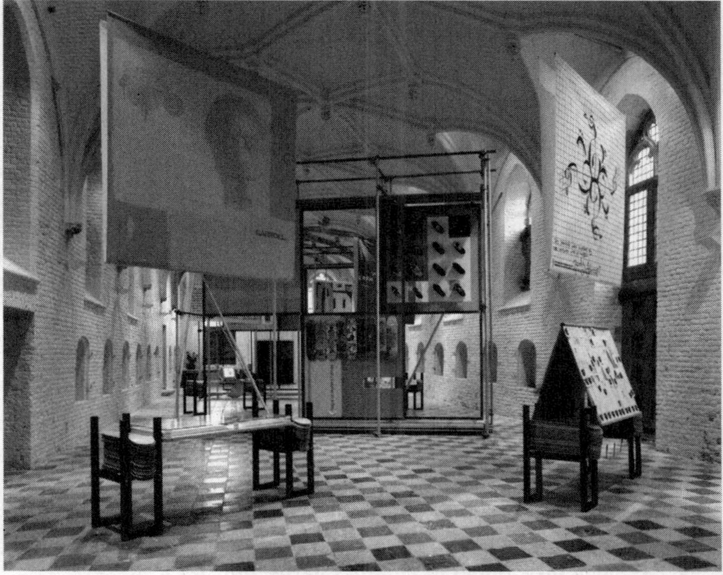

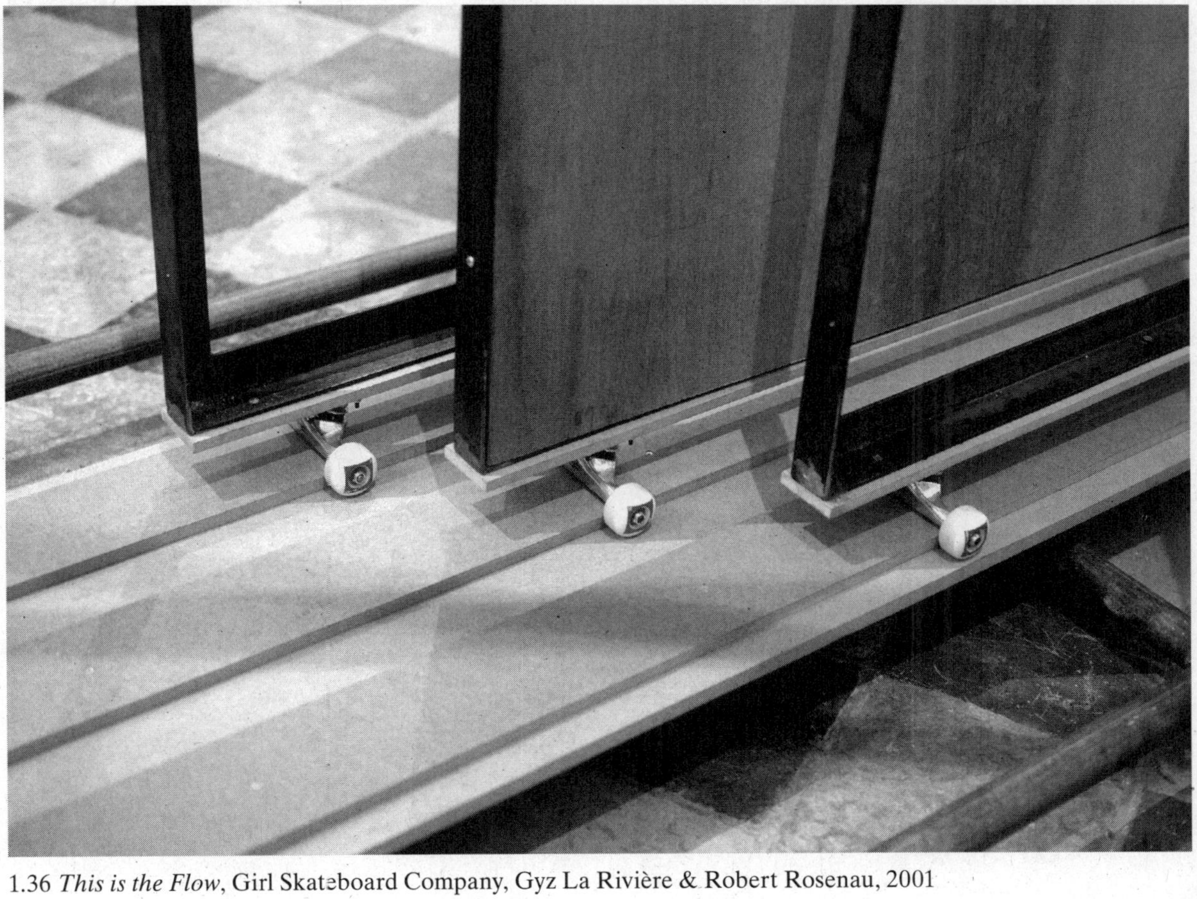

1.35 *This is the Flow*, Girl Skateboard Company, Gyz La Rivière & Robert Rosenau, 2001

1.36 *This is the Flow*, Girl Skateboard Company, Gyz La Rivière & Robert Rosenau, 2001

instructor analyses the problem: in this perfect gated community everything is under control and every danger is banished. With neither aspirations nor threats the residents of this paradise sink into inertia and apathy. So he organizes a few criminal incidents, and the resort comes to life: its residents enthusiastically start an amateur film club and begin making clandestine porn films, prescription medicines make way for recreational drugs *and* they start playing tennis again.

Ballard perhaps could not have predicted that, with the attacks on Bali and Sharm-el-Sheikh, danger would come to haunt holiday resorts in the form of the threat of terrorism.[7] But he did observe that, once liberated by individualization and secularization and unfettered by traditional social connections, we live not only without worries, but also without a sense of purpose. In the Western world we are charged with no other mission than self-development and pleasure. In this situation, leisure time has taken on tremendous significance. For the *homo ludens* of the twenty-first century, play is a serious matter, for it is only in his free time that he can give meaning to his life and, through his individual choices, show who he really is. Recent decades have seen an investment of money, energy and time on a massive scale in the leisure industry, often under the inspiring leadership of architects,

[7] The attack on Bali in 2002 resulted in 22 deaths and the attack on the Egyptian holiday resort Sharm-el-Sheikh in 2005 cost 88 people their lives.

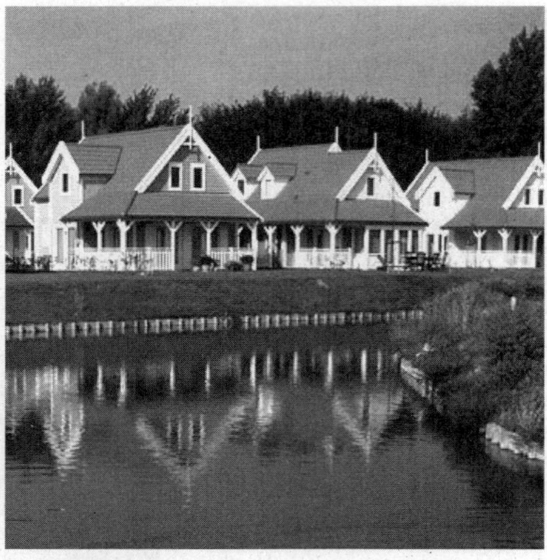

1.37 *AquaStaete*, Zeelenberg Architecture

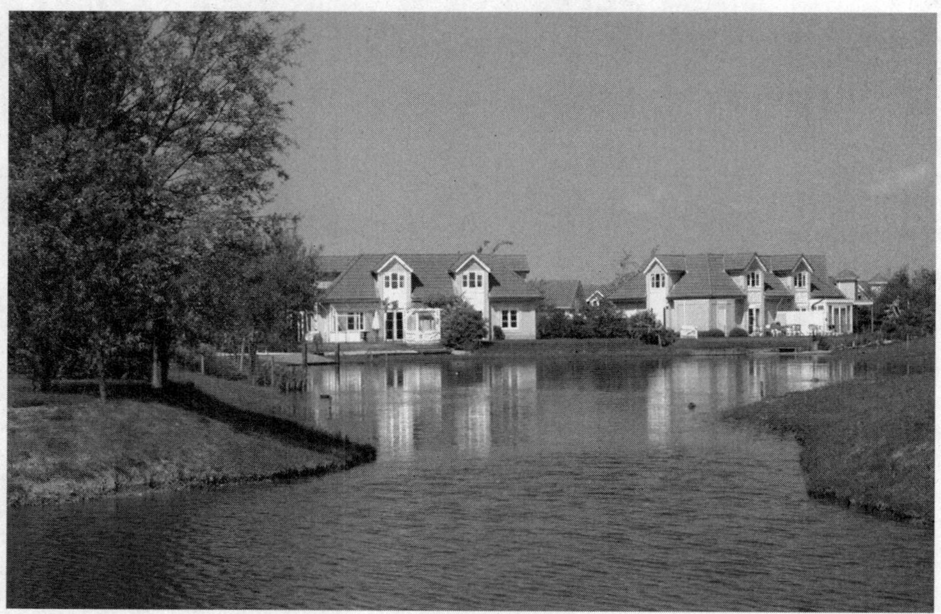

1.38 *Noordzee Residence 'De Banjaard'*, Zeelenberg Architecture

accountants and property developers who have rapidly filled the gap in the leisure market.

In the Netherlands this development is particularly apparent in the south-western coastal province of Zeeland, which has been transformed in recent years from an agricultural area into 'the Florida of the Netherlands'. Thousands of holiday homes have sprung up along the coast and in the dunes. The supply of organized activities grows daily, and when the baby-boom generation takes early retirement it can live here in a permanent state of recreation. As Middelburg, the city where De Vleeshal is located, is the capital of Zeeland, it was self-evident to organize an exhibition about the development of the leisure society.

The Leisure Society was about the strict organization of our leisure time; about Zeeland as a pleasure zone; about second homes as an investment, temple and project; and about how the leisure industry capitalizes on our mass desire to be ourselves. The exhibition was organized to coincide with the broadcast of Bregtje van der Haak's television documentary *De Vrijetijdszone* ('The Leisure Time Zone') on the day of the opening.

The Leisure Society is an example of how the museum as a 'space for ideas' can reflect upon

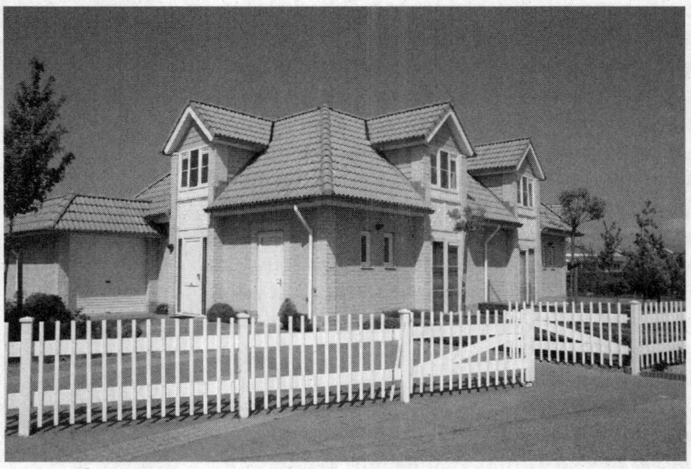

1.39 *Noordzee Residence 'De Banjaard'*, Zeelenberg Architecture

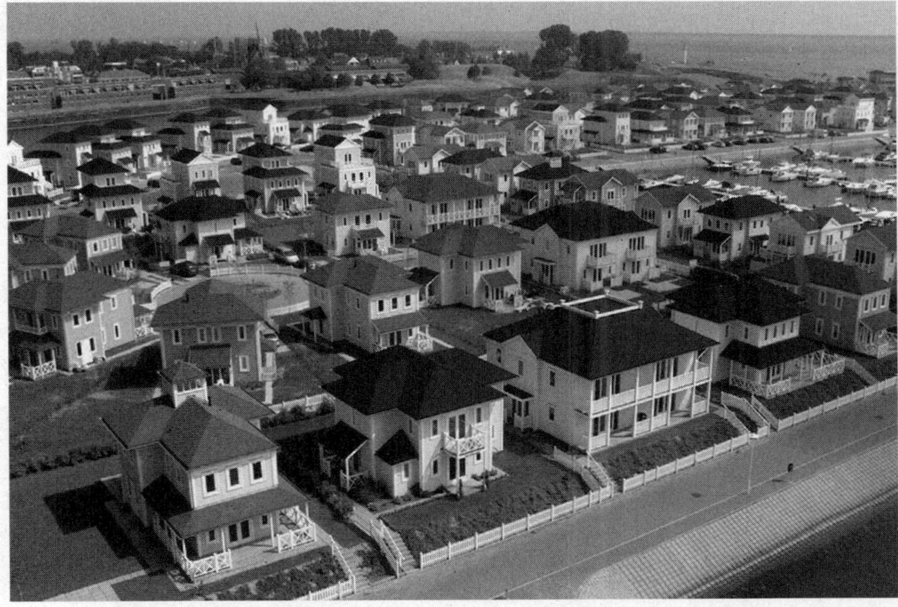

1.40 *Cape Helius*, Zeelenberg Architecture

social developments and do so in the first place from within art itself. The exhibition included paintings by Tim Stoner, which make tangible the ambiguity of the rapid emergence of the leisure industry. Drawing upon elements from popular culture and the history of painting, Stoner's images mirror our everyday (and permanently frustrated) desires for higher social status and an ultimate state of sedated sunlit perfection. His faceless figures – portrayed in a heavenly world without work or worries – call to mind the model families and blissful couples in holiday brochures. But by altering the perception of depth, detail and light in his images, Stoner presents a distorted reflection of our idealized image of free time. He has consciously chosen painting as a medium, as is clear from an interview with him: '[for me,] in this image-overloaded society, the reason to make a painting is perhaps to slow it all down a bit, not just throwing another transparent media image back into the pool of that information circuit, but making something that has a different integrity, a different speed, a different pace and a different surface' (see pages 107-113).

Art can, in other words, reflect in a unique way upon social developments. However, the context in which Stoner's work was exhibited here was not one of art history, but a mix of sociology, economy, literature and architecture. The first

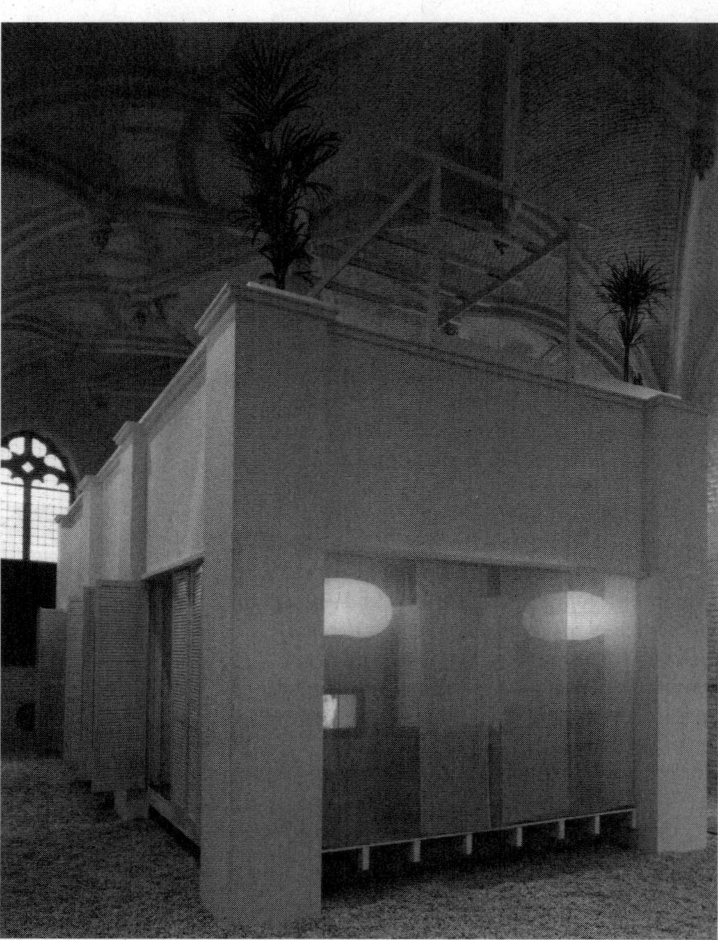

1.41 *The Leisure Society*, Tim Stoner and Zeelenberg Architecture, 2001

part of the exhibition, preceding Stoner's work, comprised a space containing statistics about the emergence of the leisure industry interspersed with visual material about future holiday sites. This section also contained quotations from sources ranging from J.G. Ballard and Arthur Rimbaud to scientific studies on the subject. In order to explore *The Leisure Society* not only from without (through art) but also from within, a pavilion was installed as a model second home. The pavilion was designed by Zeelenberg Architectuur, a practice responsible for many recent holiday developments in Zeeland, and incorporated as many popular elements from today's holiday architectural typology as possible, such as white pebbles, a bar, a roof terrace and French windows. On a sofa in front of the television, visitors could watch Van der Haak's documentary *De Vrijetijdszone*.

In general, museums do not often reflect directly upon social developments such as the changing significance of free time. The art history perspective is a barrier to such an inquiry. Museums often base thematic exhibitions on subjects derived from art history, which sets up certain limitations. The art history perspective probably has something to say about the leisure society, but it is not obvious what that would be. Furthermore, museums usually place art that has

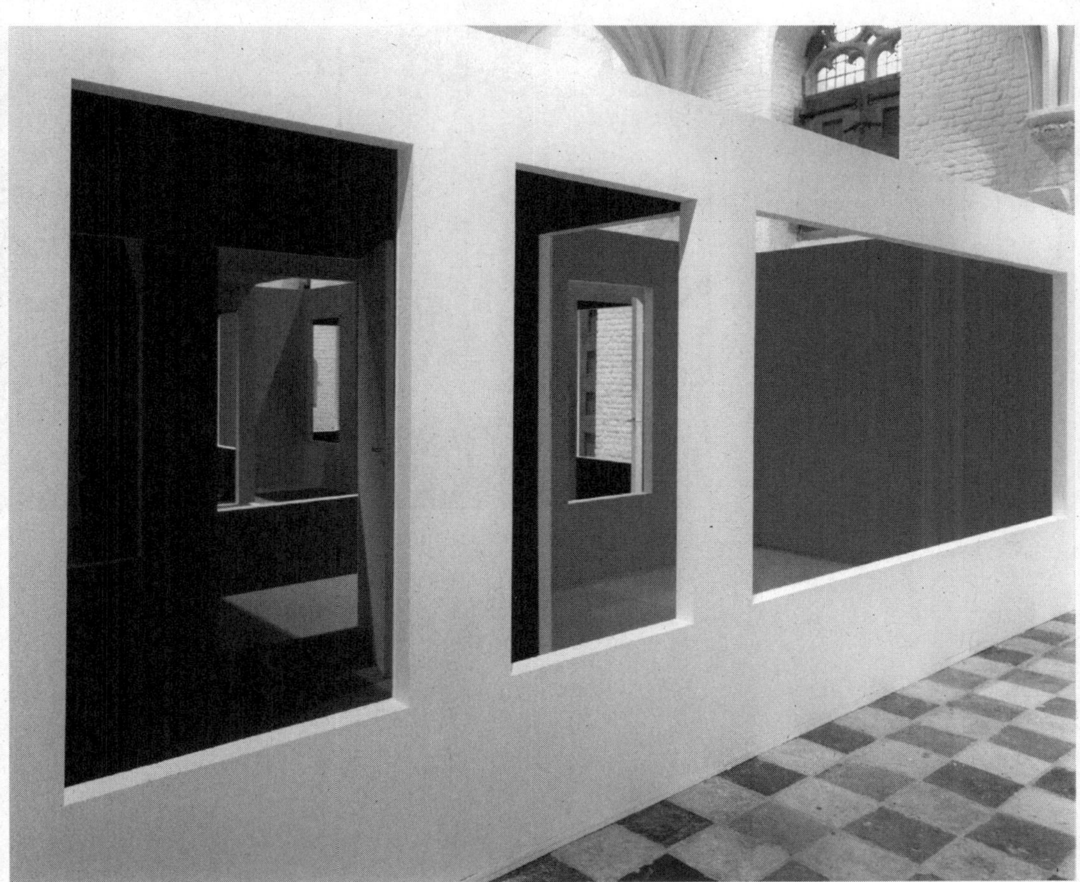

1.42 Krijn de Koning, *Beeld voor De Vleeshal*, 2000

a marked relationship to social issues in an art history framework so that, for example, an artist who is critical of globalization will be situated within a tradition of socially critical artists. They do not enter into the nature of the critique – that is down to the artist.

Museums that do seek to pursue a particular relationship with social issues nonetheless often end up in an uncomfortable dilemma, because the art history perspective offers them an insufficiently interpretive framework. The almost impenetrable jargon that one encounters in most museums is, in my opinion, often the result of their frantic attempts to extricate themselves from this dilemma. Museums can escape from this situation by unbridling themselves as much as possible from the art history perspective and by conceiving the museum as a 'space for ideas'. In this way a subject like the leisure society can be viewed from a variety of intellectual perspectives, and not exclusively through the visual arts. By following this path, it is easier for the museum to offer insights into and comment upon current social developments.

The group exhibition *Back from School*, curated by Edwin Carels, illustrates this further. The show was a reaction to the Dutroux affair in Belgium.[8] Central to the exhibition was the idea

[8] In 1995 and 1996 Marc Dutroux abducted, tortured and abused six girls aged 8 to 19, four of whom he murdered. His trial highlighted several blunders by the police and the public prosecutor that brought discredit upon the Belgian judicial system. The Dutroux affair caused widespread shock in Belgium and far beyond its borders.

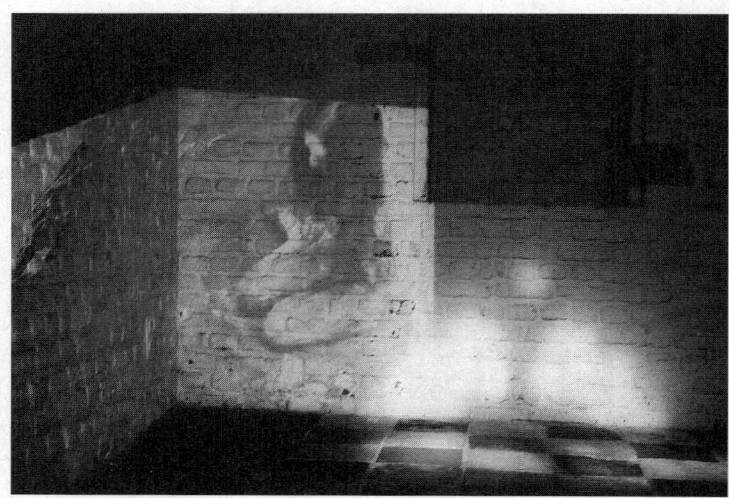

1.43 *Terug van School*, installation view, 2000

of innocence that adults project onto children. In
the essay 'Unguarded Moments' in this volume,
Carels writes: 'It is an ever-more vicious circle as
the child turns to the media to construct his or her
self-image, while the media relies more heavily
upon canonical images of childlike innocence.
In Belgium in recent years the idea of children's
innocence has been cultivated to an especially
high degree while equally strong fantasies have
been created about how this innocence might be
lost' (see pages 193-218).

A retort to such ideas was provided by *Back
from School*. The exhibition's invitation card
bore a photograph of a small boy sitting on the
ground playing with an overturned wheelbarrow.
The boy, who looks rather sad and neglected, is
Marc Dutroux. The central work in the exhibition,
Marijke van Warmerdam's *Handstand* from 1992,
created a similar kind of confrontation. In this
film loop we see a young woman in a white dress
repeatedly perform a handstand during which
her dress slips down to reveal her white knickers.
Van Warmerdam made this work long before the
Dutroux affair, but by presenting the work in the
context of this exhibition, Carels made clear to
what extent our vision can be coloured. 'Whoever
looks for it can find a dubious hidden meaning
in the most innocent of works,' writes Carels in
'Unguarded Moments'. The fashion designer

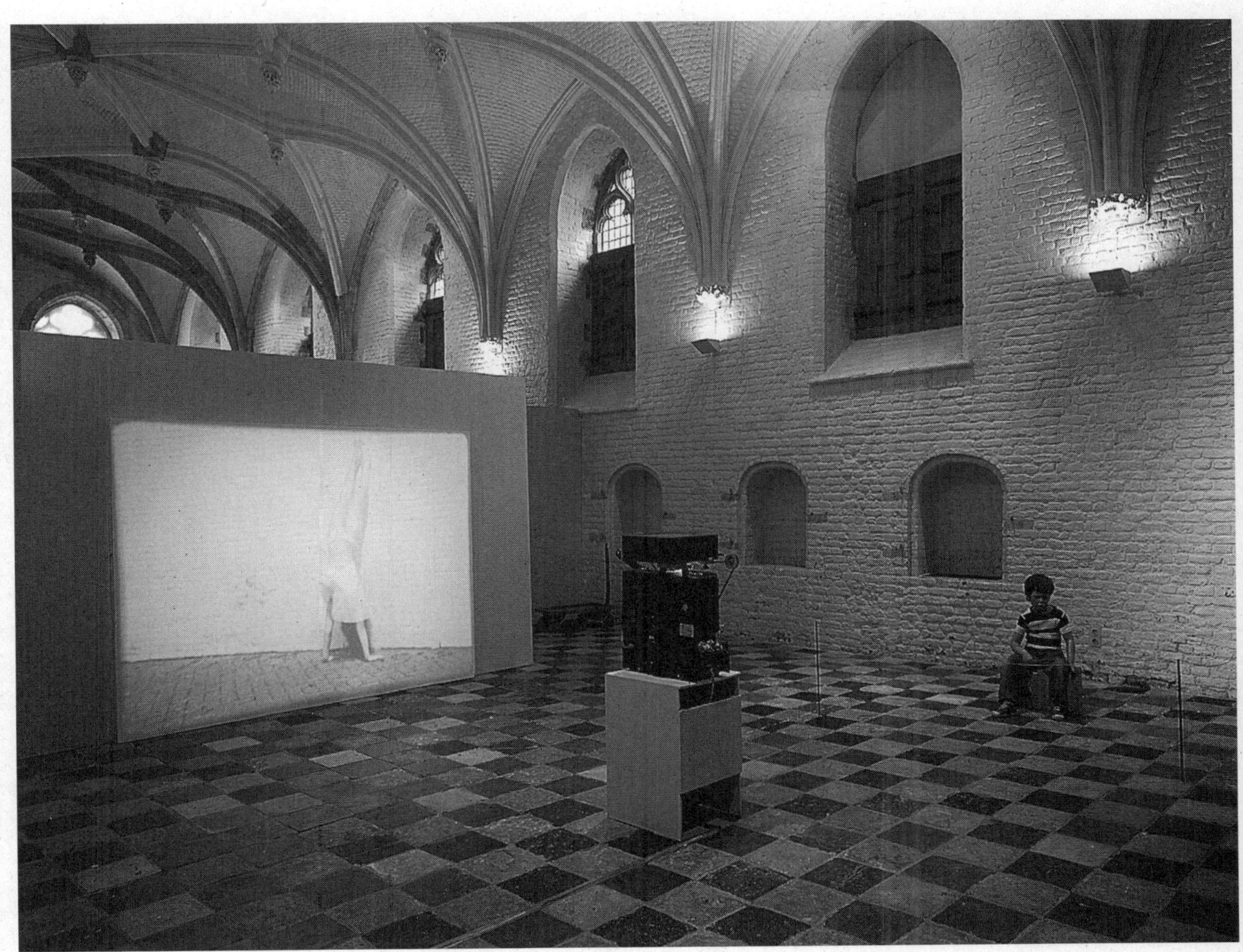

1.44 *Terug van School*, installation view, 2000

Martin Margiela seemed more than aware of this. The exhibition included several outfits in which he had enlarged dolls' clothing to adult sizes.

Another exhibition in this series was *Safe Haven... or the Aesthetics of Safety*, curated by Guus Beumer. After 11 September and the assassination of the Dutch right-wing populist politician Pim Fortuyn in 2002, Dutch politics and society was ruled by an obsession with security. *Safe Haven... or the Aesthetics of Safety* examined our desire for security (and our fear of security breaches) from an aesthetic perspective. The exhibition employed three scenarios to place the current aesthetics of security within a historical context, making connections between interior design and the world view that hides behind it.

The first scenario, with velvet-upholstered furniture and silk wallpaper, was reminiscent of a nineteenth-century salon. This interior exuded the womb-like warmth that protects the middle-class family from the dangerous and undisciplined world outside. The second scenario referred to the 1930s. The design of this interior, consisting of a smooth tiled wall and steel and glass furniture, exuded the belief in progress typical of the modernist movement's International Style. This movement longed for a new world created by modern man. The third scenario illustrated

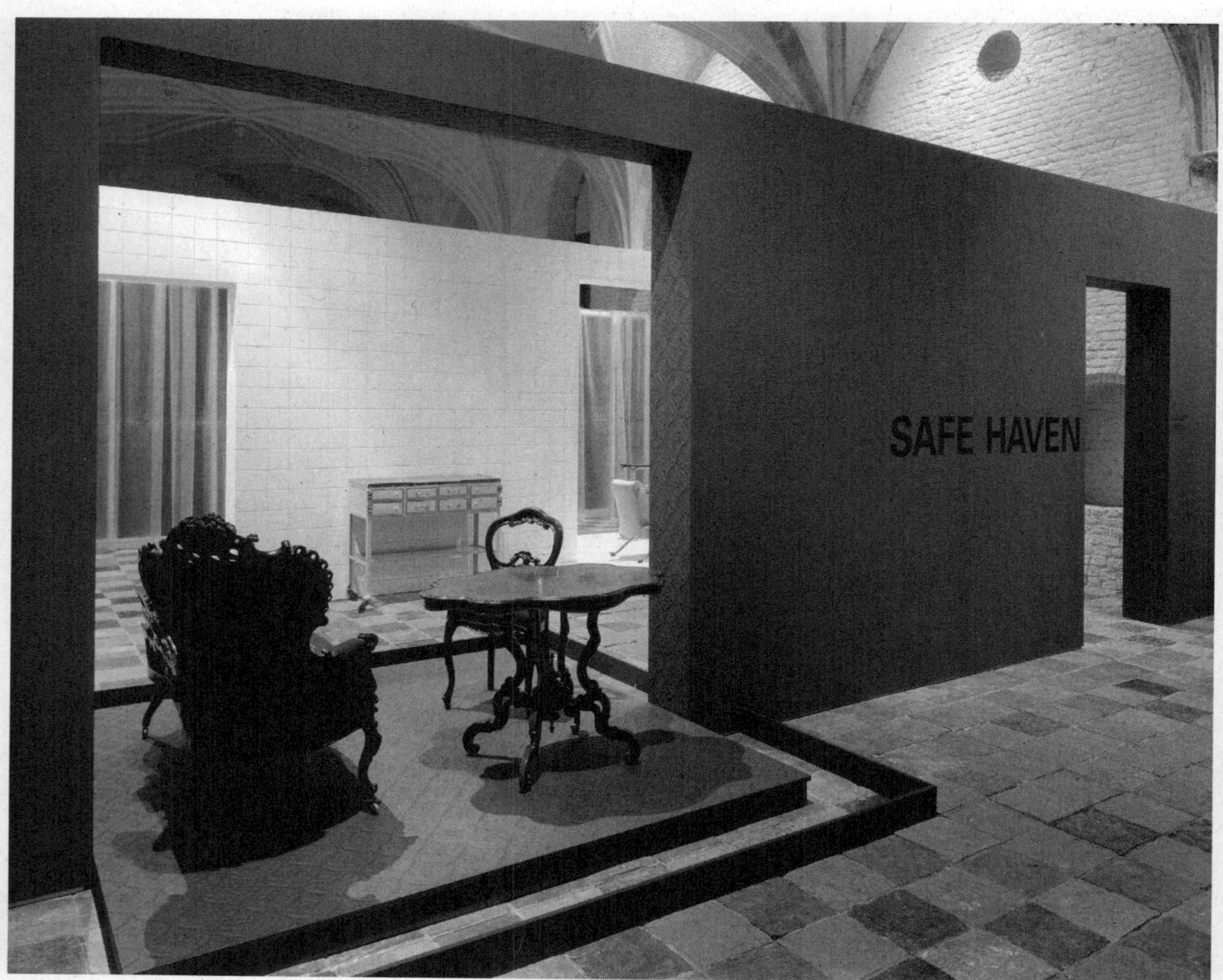

1.45 *Safe Haven... or the Aesthetics of Safety*, installation view, 2003

the aesthetic of safety at the beginning of the twenty-first century. The exterior and interior were fused, both by the use of glass and by the extensive use of stainless steel. The glass not only reduced the difference between private and public space but also increased visibility and thus the feeling of control. We associate the use of stainless steel in kitchens and bathrooms with luxury, but in the public realm the same material is often employed precisely because it is both hygienic and vandal-proof.

The pieces of furniture exhibited in *Safe Haven... or the Aesthetics of Safety* were all loaned from the decorative arts collections of museums. Usually these objects are used to illustrate the history of design and how the availability of new materials and techniques has led to new forms. But *Safe Haven... or the Aesthetics of Safety* showed that the same items could also be enlisted to reflect upon issues with a greater social urgency.

An exhibition like *Back from School* is essentially a visual essay that poses the question of how we can perceive the appalling events around the Dutroux affair and thus – ultimately – reach a conclusion about how to deal with them. Like *The Leisure Society* and *Safe Haven... or the Aesthetics of Safety* it attempted to offer insights

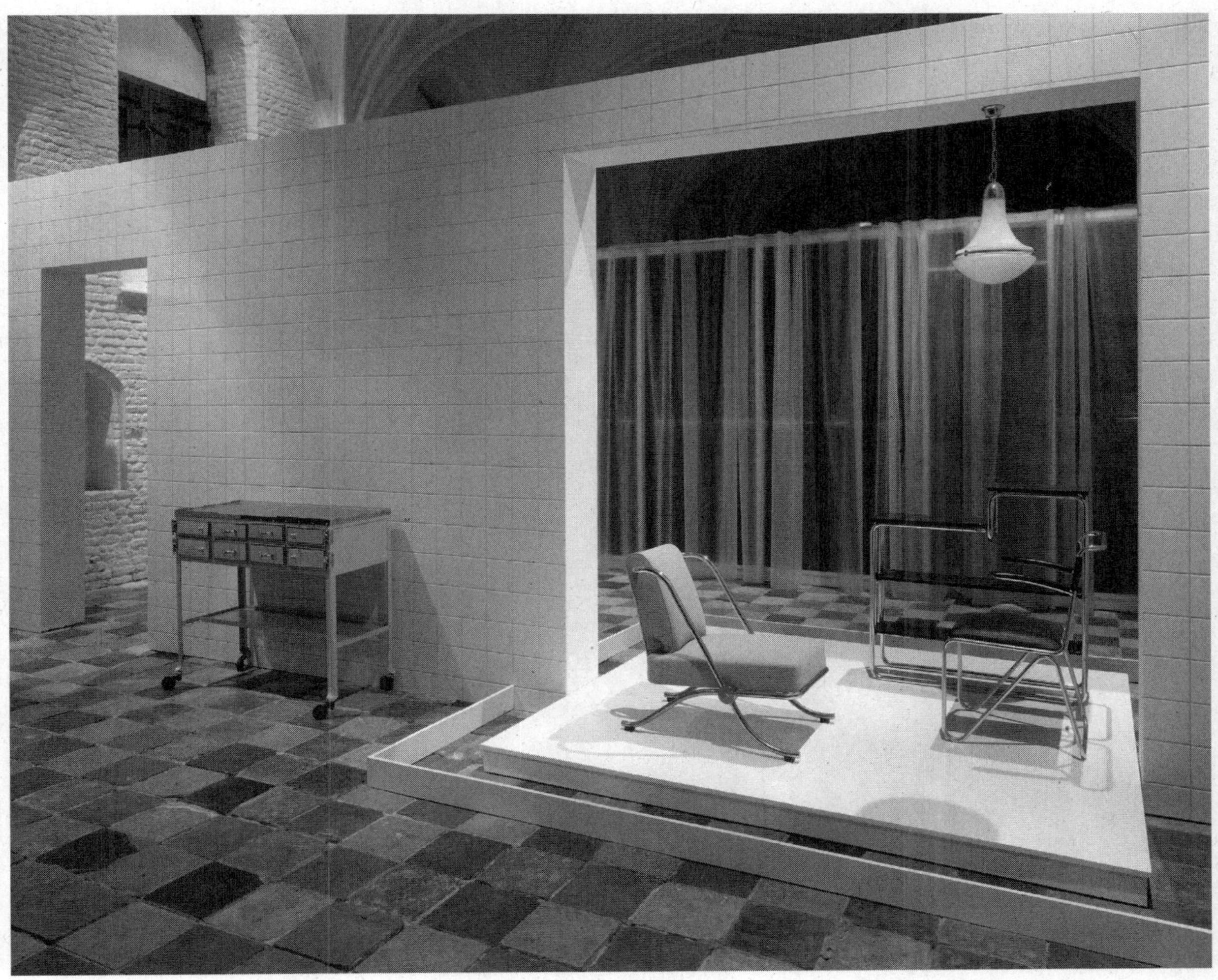

1.46 *Safe Haven... or the Aesthetics of Safety*, installation view, 2003

into fundamental processes of social change. Of course such ideas and insights are also formulated elsewhere, for example within scientific disciplines and the media. The latter are as capable of providing nuanced comment as they are of producing sensationalism, but the museum's privileged stance for reflecting upon society is self-evident.

Museums, by their very nature, offer more time and space for reflection. Things can simply be, without having to justify their existence by the amount of attention they attract, as is the case with the mass media. Moreover art – and therefore the museum – has greater freedom than science, which is bound by laws. Museums, by contrast, are almost as value-free as science. Outspoken political or religious sentiments need not play a role in museums. Thanks to these qualities, the museum as a 'space for ideas' can also be interpreted as a special form of the public realm in which debate and reflection upon society can take place on its own terms. In its role as a 'space for ideas' the museum automatically legitimates its special place within society.

The project *New Symbols for the Netherlands* from 2005 is perhaps the most far-reaching example of this. The project was inspired by an article by Anna Tilroe in a Dutch national

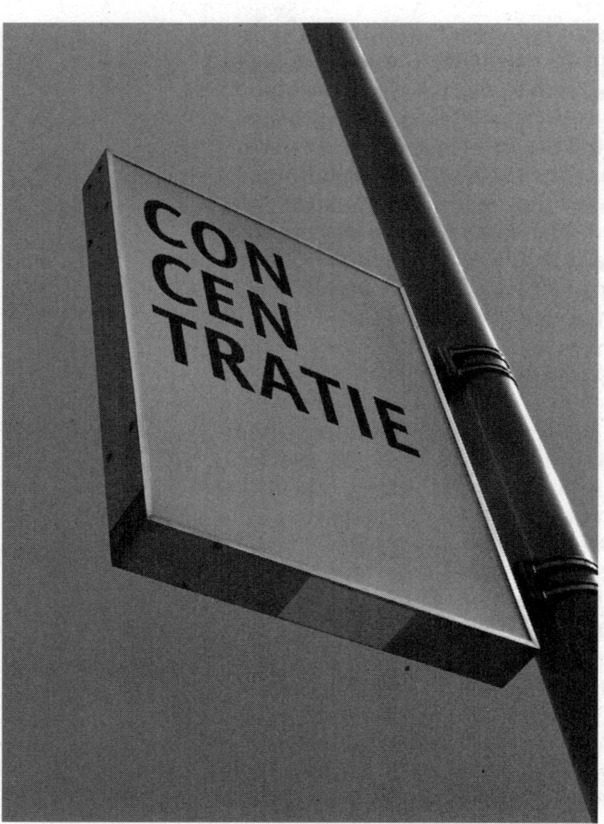

1.47 Annelys de Vet, *Concentratie* ('Concentration'), 2005

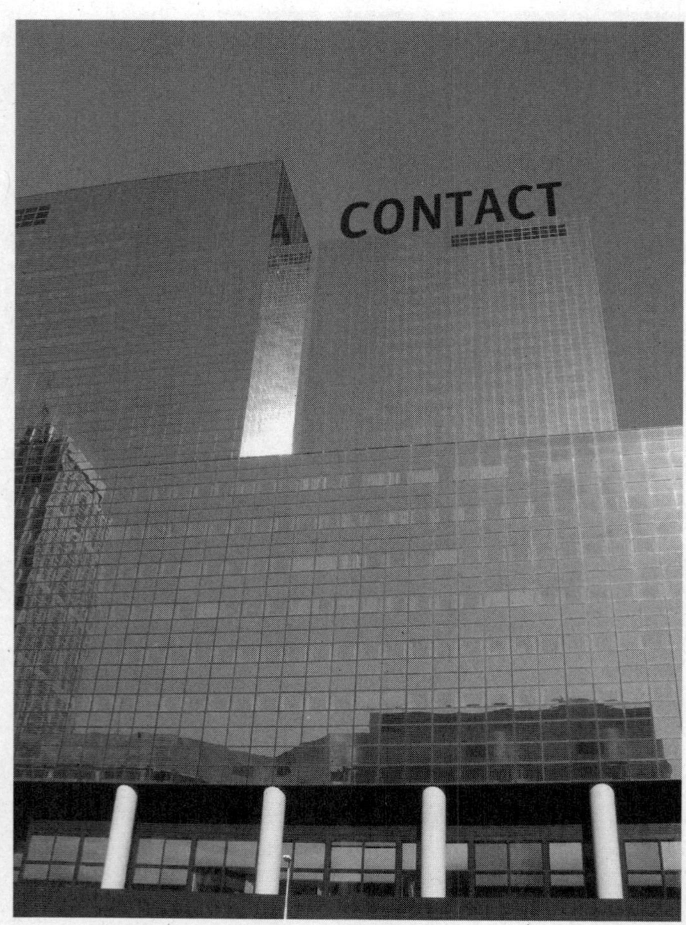

1.48 Annelys de Vet, *Contact*, 2005

newspaper in which, following the attacks of 11 September and the murders of Pim Fortuyn and the filmmaker Theo van Gogh, she identified the need for, 'weighty, lasting symbols that can stand up to the tyrannical symbolism of terror'.[9] According to Tilroe, art is still capable of producing symbols, and so artists should tackle the question of how they can develop symbols that are 'authentic, meaningful and inspirational'. But in order to do so, art must once again dare to deal with values – values connected with the sovereignty of the individual and the freedom and solidarity of society.

Tilroe's appeal transcended the traditional distrust of symbols and fear of moral kitsch. This was all the more remarkable because the Dutch liberal left had initially reacted with such derision to the debate about norms and values previously initiated by the Christian-Democrat prime minister, Jan Peter Balkenende. What struck me most was that the article confronted my own reflex to distrust symbols and to perceive the discussion about values as redundant. But does the time we are living in not call for a different stance? After all, it is no longer possible to speak of a shared idea about a common future, and Tilroe was right in asserting that the influence of globalization, other cultures and religions and a series of attacks make us long for symbols that express those

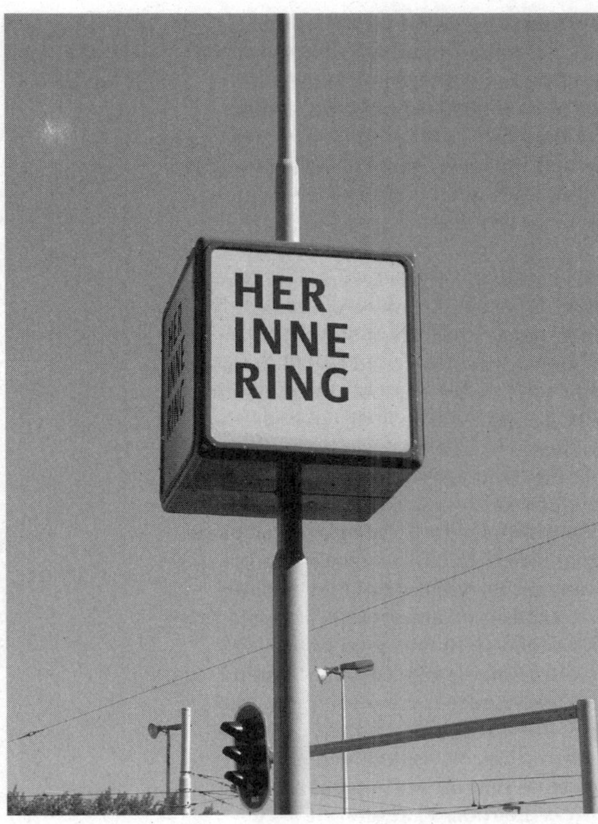

1.49 Annelys de Vet, *Herinnering* ('Memory'), 2005

[9] Anna Tilroe, 'Het grote gemis'. *NRC Handelsblad*, Rotterdam, 17 December 2004.

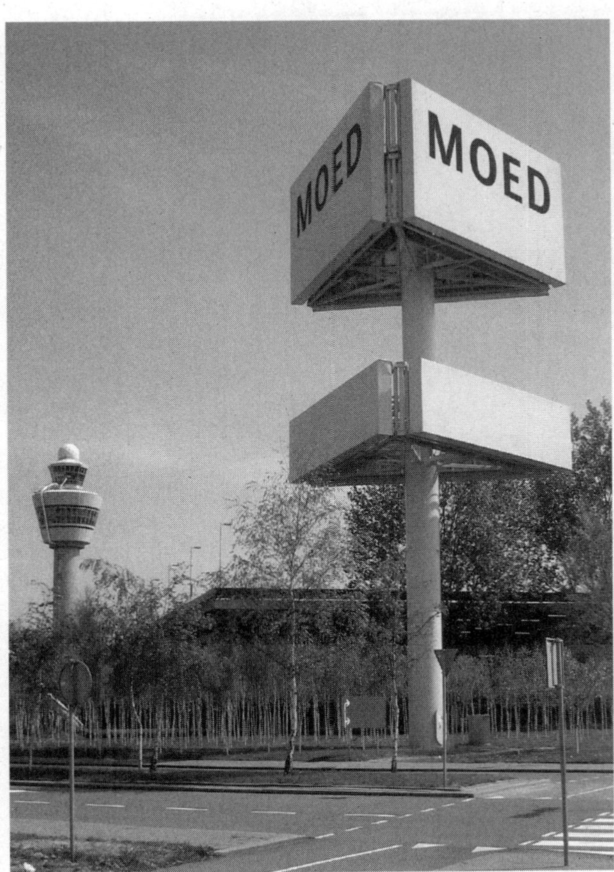

1.50 Annelys de Vet, *Moed* ('Courage'), 2005

things that bind us together. Can artists create new symbols that not only make visible the moral void but also help to fill it? Can we bridge the gap between individual freedom – and thus the autonomy of the artist – and communal values? Formulated in these terms, there is much at stake for art, for could this not increase art's role and significance in society?

Reason enough to wish to respond to Tilroe's call to arms with the exhibition *New Symbols for the Netherlands*. A painter, a playwright, a graphic designer, an architect and two filmmakers were invited to create new symbols for the Netherlands. This was prefaced by two discussions about what values these symbols should express. The discussions were collected in the publication *Nieuwe symbolen voor Nederland*, together with five essays, including Anna Tilroe's original piece. The participants in the discussions were united in the opinion that the creation of new symbols for the Netherlands is an almost unachievable ambition. And the values these symbols were to express – altruism, dignity, courage, contact, curiosity and resistance – were formulated only after considerable vacillation and not without a degree of reservation. Nevertheless the conversations demonstrate that the museum as a 'space for ideas' offers the room to deliberate such confronting issues.

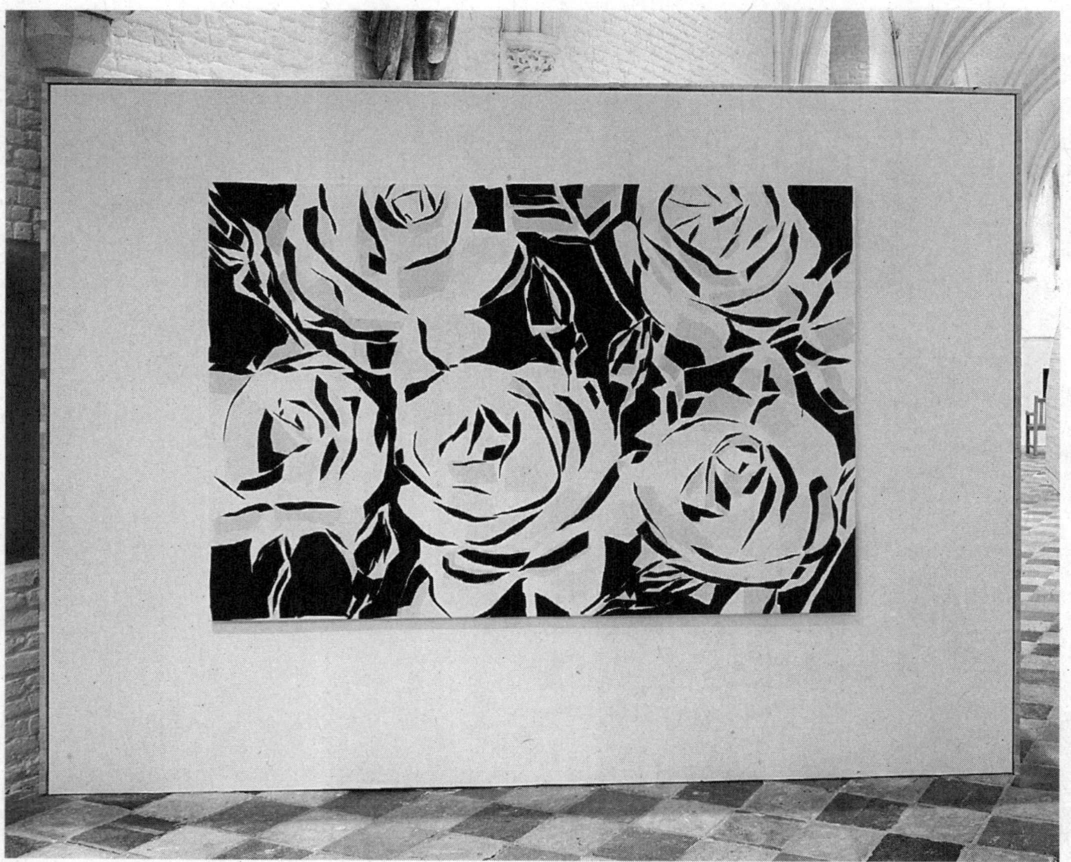

1.51 Marjolein Rothman, *Roses* in: *Nieuwe symbolen voor Nederland*, installation view, 2005

Art does not always require explication, whether from an art history perspective or any other perspective, for that matter. A work of art can manifest itself without any context to add anything of essence to it. As a child I was fascinated by the abstract works of Piet Mondrian in the Gemeentemuseum in The Hague. Without knowing anything else about his work, I felt his desire to evoke mystical experiences with nothing more than pure colours and simple compositions. Such an experience is unique to a museum. But museums can no longer appeal solely to such experiences for their justification. The age we live in demands that museums' ambitions transcend the level of the individual experience, no matter how important and special it might be.

The inviolable position that art has long enjoyed in our culture is today less self-evident, and so the authority of the museum has come under pressure. Museums themselves are partly to blame for this. In formulating their mission, they have too long restricted themselves to truisms and vague abstractions or to a definition of their core remit. But politics has also played its part by legitimizing art and museums by means of secondary arguments, such as their value as instruments for social integration or as an economic engine. Meanwhile, for many people art – and visiting museums – is nothing more than

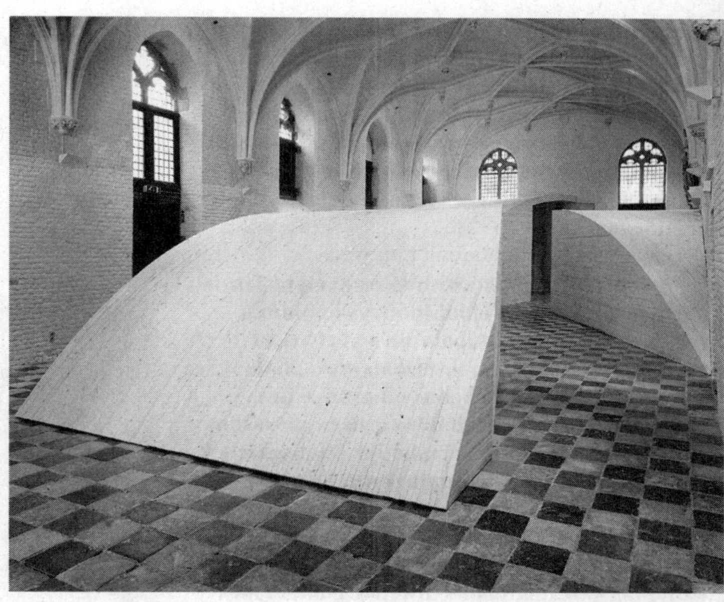

1.52 *Nieuwe symbolen voor Nederland*, installation view Alex van de Beld, 2005

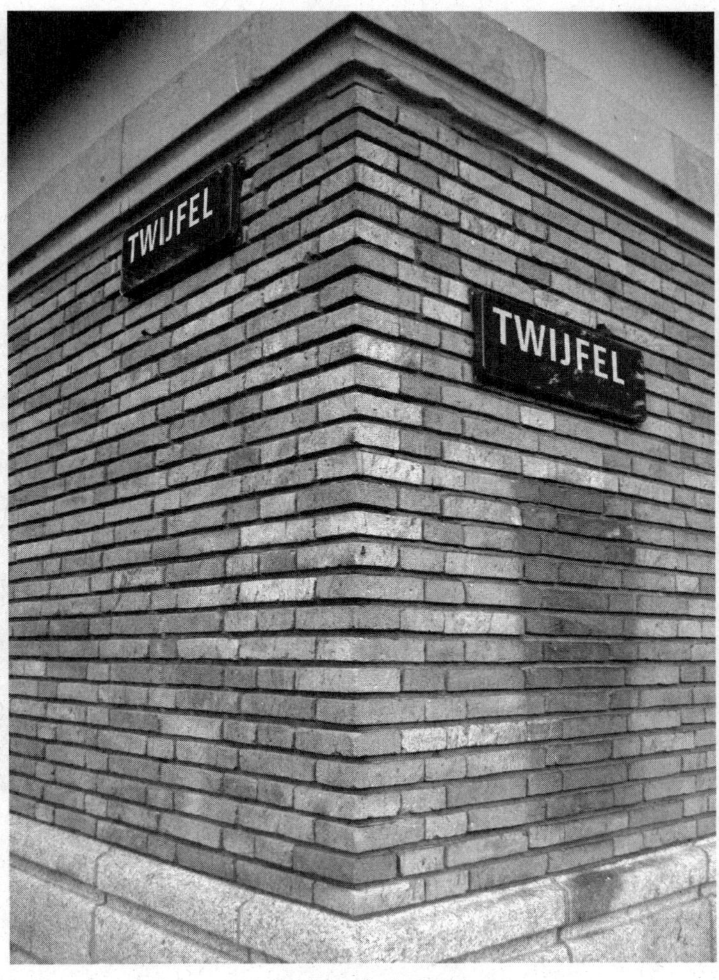

1.53 Annelys de Vet, *Twijfel* ('Doubt'), 2005

another component of their lifestyle. It is therefore essential that museums reformulate the significance of art for society and thus of their own role. If museums are incapable of this, they will increasingly be unable to convince others of their worth, or worse: politicians and policy makers will do it for them.

But this does not mean that we need to completely jettison the art history perspective. Modern art (roughly from 1860 to 1980) is well served by elucidation from an art history perspective. But modern art does more than simply refer to its own history. To make this visible, museums must employ other approaches besides the art history perspective, particularly when it comes to contemporary art. Whereas the history of art up to the end of the modern period demonstrates a clear progression of successive styles and movements, contemporary art is rather more wayward. Attempts to define contemporary art from an art history perspective all too often become mired in an entirely hermetic argument so insular that it is of interest to only a few.

To my mind, museums must free themselves from this straitjacket by reformulating their ambitions. They can increase their role in society simply by making it a greater focus of their attention. By conceiving themselves as a 'space

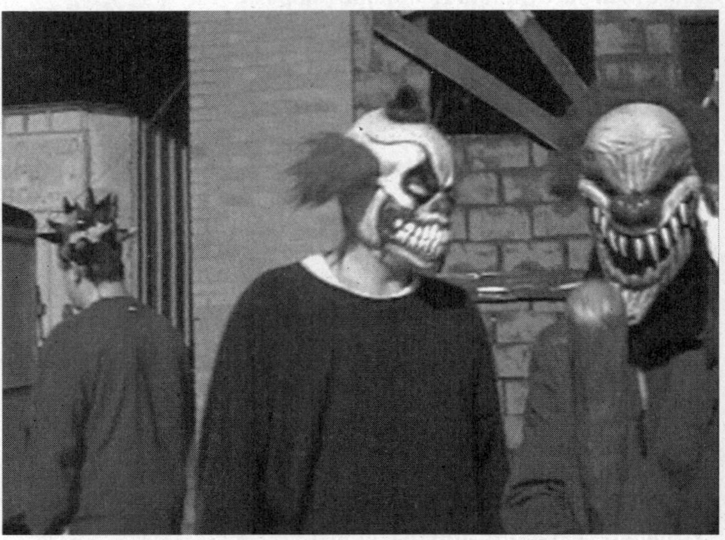

1.54 Cameron Jamie, *Spook House*, 2003

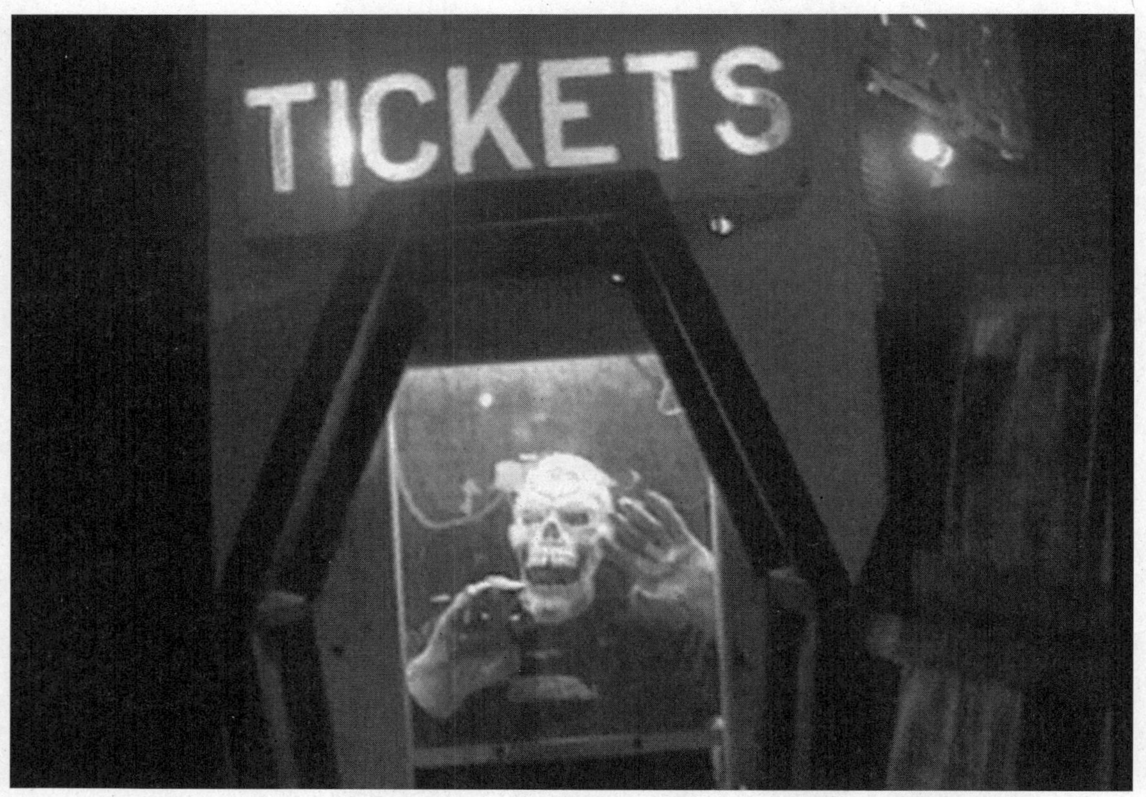

1.55 Cameron Jamie, *Spook House*, 2003

for ideas', museums can offer insights into our rapidly changing world and thus ourselves. This can be achieved by reflecting upon developments in society through art or by conceiving the museum as a privileged form of the public realm: a space for reflection upon and debate about society. But in order to do this, they must first dare to let go of the dominant perspective of art history. This requires nothing less than museums rediscovering themselves and encouraging art to do the same; for it is not only the future of museums that will be defined by their ambitions – but the future of art as well.

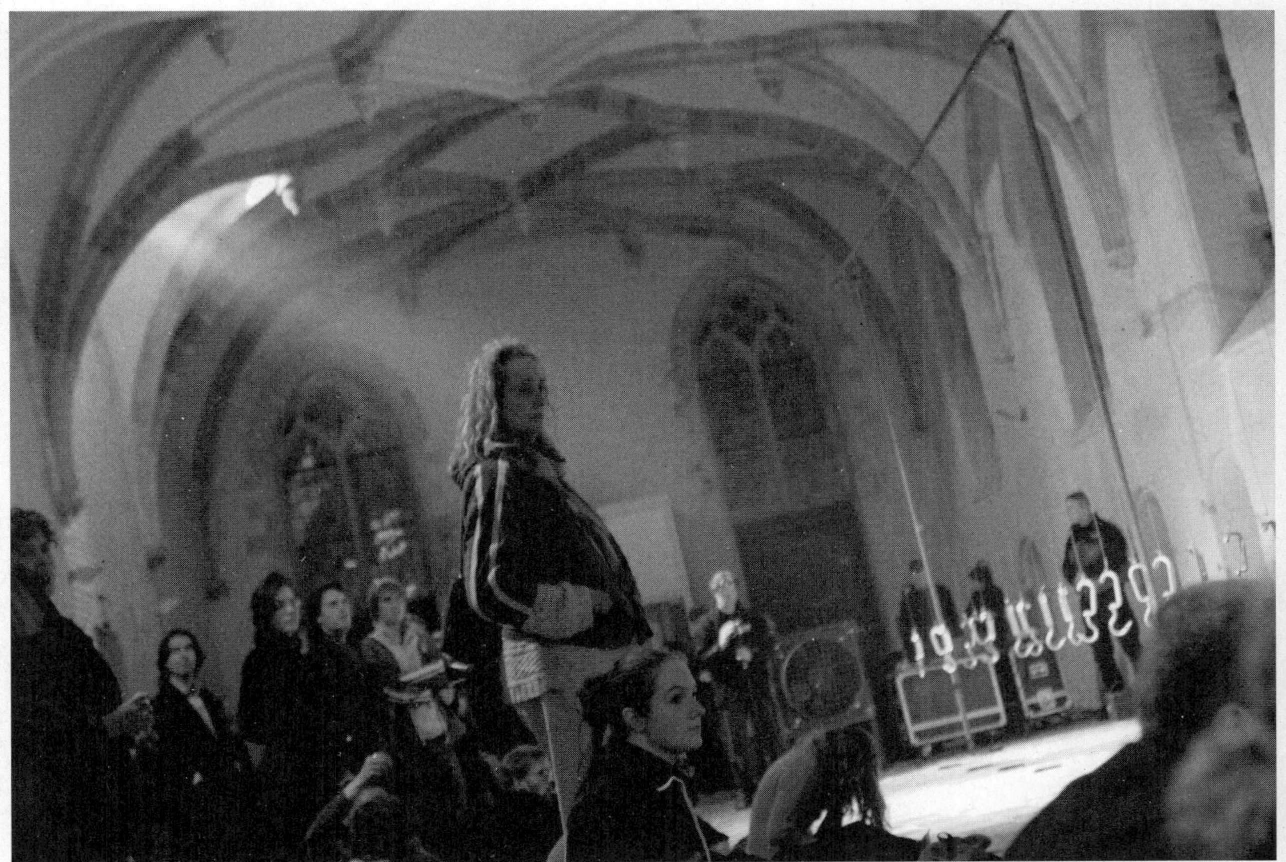

1.56 *Pan Sonic Live*, performance, April 9, 2005

Anna Tilroe: The Promise

International airports are the vista of post-industrial society. They form a world where disciplined crowds noiselessly move around in a carefully constructed ambience of orderliness, luxury and perfection. It is a world that commands respect for its underlying organizational ingenuity and mass psychological insight, and for the material subtlety of its design. Nothing may be reminiscent of the chaos, arbitrariness and dangers of the world beyond customs. Everything is under control and, at the same time, full of promise: we have left, but we have not yet arrived. Things have yet to happen; time, for the moment, is suspended.

Some thinkers call airports 'non-spaces', by which they mean generic places: waiting rooms, stations, industrial zones. The airport does indeed evoke a sense of being a grey area, a vacuum where the very concept of identity crumbles like a dried-out label. However, the term 'non-space' is not entirely appropriate. Airports such as those in Amsterdam, Shanghai and Tokyo do have character, if only because they have been thought out not in terms of the past or the present, but in terms of the future. What exactly do these visions of the future encompass? What are the concepts behind the design? Where does all this leave

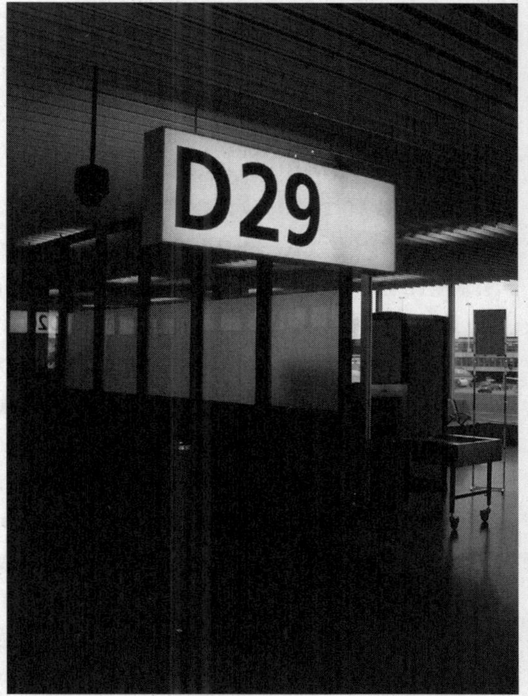

2.1 Schiphol Airport, Amsterdam, 2005

people, now and in 2015? These are questions that Brazilian artist Ana Maria Tavares must have contemplated when she decided to transform De Vleeshal into a futuristic *Airport Lounge*.

Tavares has not needed much: long stainless steel poles reaching up to the ceiling; turnstiles and crush barriers, also in stainless steel; two enormous mirrors; and two large screens onto which video images are projected. This proves sufficient to create an imposing minimalist spaciousness and an atmosphere in which the future floats as a cool, dazzling light.

This coolness is often seen in environments designed to reflect a positive orientation toward the future and great expectations of high technology. Stainless steel's unapproachable quality generates an image of discipline and inflexibility – an image, in other words, of a society governed by functionalism, and spurred on by an almost superhuman will, the will of the technological system. The resulting alienation of our existential basis was strikingly depicted by film director David Cronenberg, a few years ago, in his film *eXistenZ*. In this film a young couple rebels against a 'mind-fucking' technological invention, both of them wielding weapons made out of flesh and bone: jagged and bloody, but impressive for being purely *natural*.

2.2   Ana Maria Tavares, *Relax'o'visions*, 1998

Tavares's rebellion is not explicit. She dissects and reassembles, fully aware of the enchantment brought about by beauty and power.

Her point of departure is the notion that an airport is a marvel of rational and technical ingenuity – transcending all that is earthy, the human being included. And so she has first captured the panorama of the airport. This is largely achieved by the gigantic, slightly inclined mirror on the floor. It reflects the beautiful arched ceiling, but simultan-eously changes the ground into a dark firmament, a sensation heightened by the black and white video projection on the wall behind the mirror. It shows an empty escalator slanting upwards towards something enveloped in darkness. In the mirror this darkness deepens, becoming a hole in the ground.

Another video projection, this one in the artificial colours of a video game, provides a variant on the suggestion of endlessly moving masses. Nowwe see a path, strewn with semicir-cular stainless steel gates, seemingly leading towards an underground space. The gates look as if they are supposed to keep long rows of waiting passengers on course, but there is no system to them. We slalom aimlessly around the gates – and still there is more confusion to come. The mass

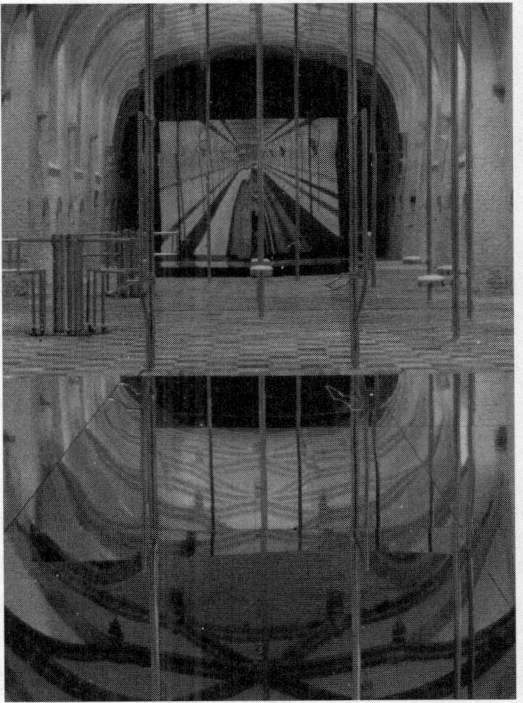

2.3   Ana Maria Tavares, *Middelburg Airport Lounge with Parede Niemeyer*, 2001

exodus towards the unknown at the end of the hall is doubled by the enormous mirror on the opposite wall. The effect is overwhelming: beginning and ending, projection and reflection, all become interwoven. It does not matter anymore what is real and what is not.

Airports, so the mirrors and projections echo, are imposing, both because of their *mise en scène* and because of their streamlined boundlessness. They reduce people to an ant colony in which each individual has a function, but where all individuality dissipates. In this sense, they are metropolises like Los Angeles or São Paulo. But international airports are also capitalist enterprises: they flourish when customers are happy, be they passengers or shopkeepers who have leased a bit of airport ground (the most expensive land on earth). And so everything possible is done to preclude feelings of alienation and panic. We all need to get into the 'lounge mood' as soon as possible.

The lounge is a mood maker. Here, people placidly sit passing the time, reading, looking around, drinking, but without feeling pressured to interact socially as they would do in a pub. It is a kind of meditation centre, where you can be alone amidst others: a luxury filched from a hectic existence geared towards mobility. In this sense,

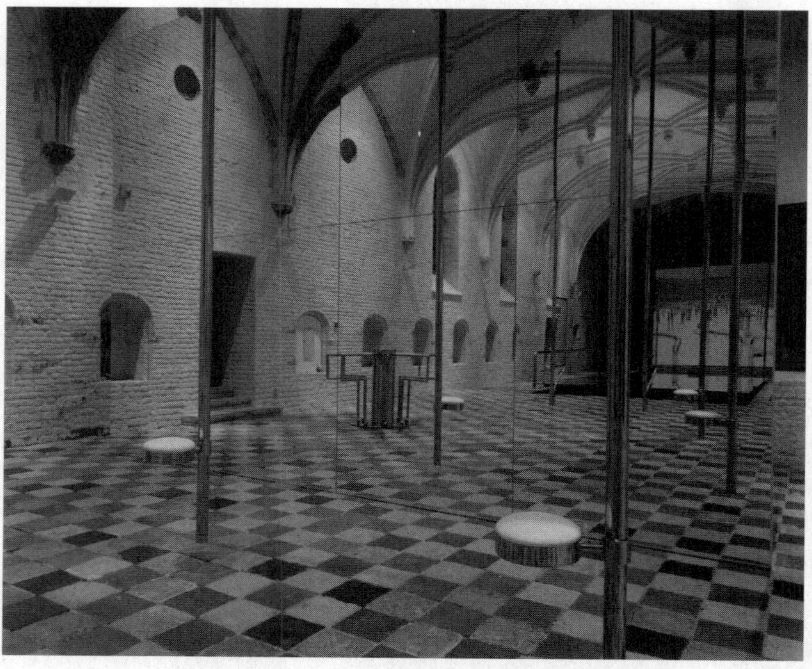

2.4   Ana Maria Tavares, *Middelburg Airport Lounge with Parede Niemeyer*, 2001

the lounge has an regulating effect. Panic subsides; reins are loosened; things are left to run their course. This is the perfect mood for the air traveller, the mood the airport's design aims for. It moves us to relinquish control over our lives and to step out of time: *ready for take off.*

In De Vleeshal the lounge consists of eight stainless steel poles that stand in a semicircle, reaching the ceiling. Each pole has a round seat fixed to it, covered by a fat white pillow. There is some distance between the poles, so you don't have to chat with your neighbour if you don't want to. You can intensify your self-chosen isolation by slipping on one of the cordless headphones hanging invitingly from a barrier. They emit groovy music from the 1950s and 1960s, alternated with typical airport sounds: the tranquil buzz of a shopping, strolling crowd; the professional, yet friendly messages transmitted over the PA system; the faraway rumble of airplanes landing and taking off.

Listening to these agreeable sounds, your 'lonely planet' becomes a pleasant vantage point from which to look up at the heavenly arch overhead and, reflected in the mirror beneath your feet, the gently moving escalator and the slalom manoeuvres of the gates. Everything exudes perfection and a feel for the grandiose.

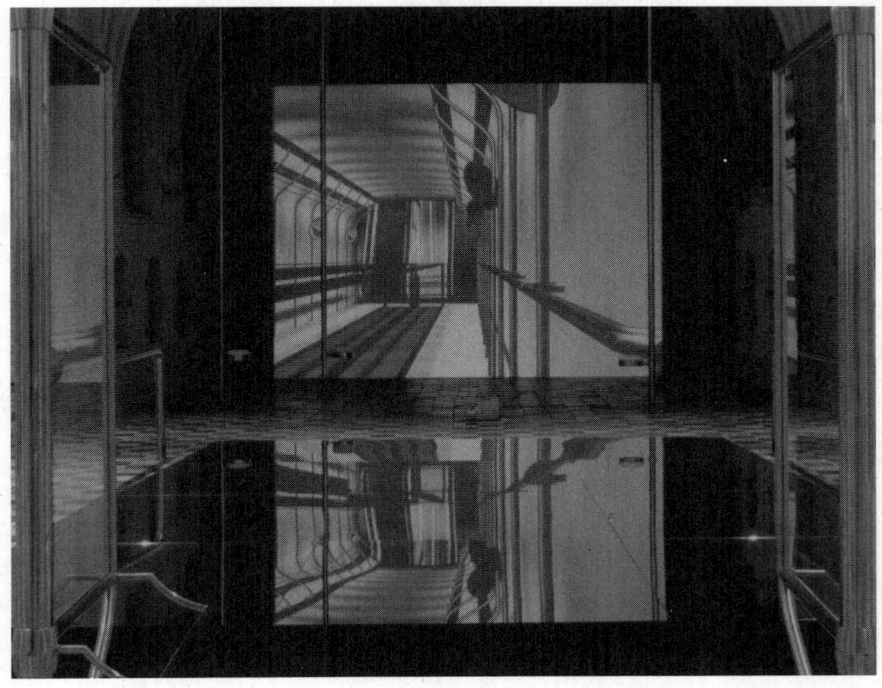

2.5   Ana Maria Tavares, *Middelburg Airport Lounge with Parede Niemeyer*, 2001

As an airport, this place – indeed the world itself – has taken on a new, architectural style, that inspires awe for its openness, lightness and monumental seclusion.

In this frame of mind you readily agree with what the young Dutch pair of architects Ben van Berkel and Caroline Bos write in their book *Move*: 'In architecture, the "Made in Heaven" effect is expressed most purely in perfectionist buildings that give you a rich feeling and cause you to continuously gaze upward and to the side. To enter their hollow bodies with your own body enlightens you. To walk through them is to walk through a painting: you see what you choose to see; your gaze swerves and orients you through colour, shininess, light, figuration and sensation.'

The world as architecture, architecture as painting: this vision of the planet as a work of art is truly majestic. And yet something is up.

Is it because of the long, smoothly polished fences on either side of the horizontal mirror that a sense of restriction seems to loom over this man-made paradise? Surely their sole purpose is to prevent you from leaning too far forward and drowning in the mirrored ceiling. Or is it the clustered turnstiles, standing among the seating poles like fan-shaped sculptures? You look around. There are the poles with the swing gates. For the moment they permit free passage, but a sense of control lingers about them – and all of a sudden 'control' is palpable everywhere. Control: aimed at imposing discipline and compliance. Control: designed to steer us, mentally, in the right direction – towards the shopping mall and the lounge. Control: to prevent inevitable rebellion. Control: as 'Evil' lurks everywhere!

No paradise without fences: we knew it. But we needed this artwork of Tavares's to see it more sharply than ever.

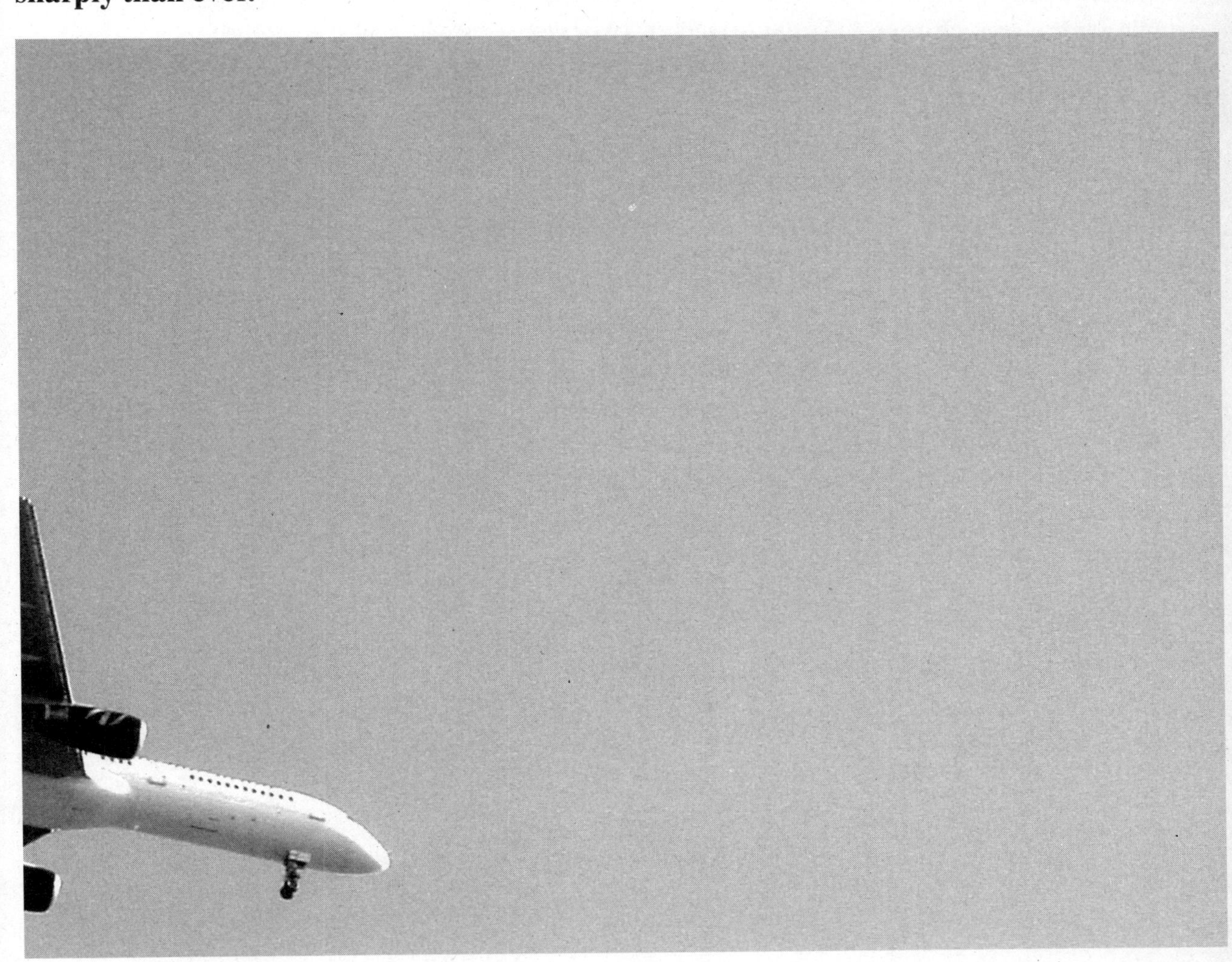

2.6   From LAX Airport, 2001

Bas Heijne: Reality

When I was young, any artist worth his or her salt still simply committed suicide. At secondary school in Badhoevedorp in the mid-1970s, literature taught us that life was basically unbearable. And it was art that forged a path to the emergency exit – time and time again. In French lessons, Rimbaud advocated the deliberate 'deregulation' of all the senses, synesthesia: *I is somebody else*. During tedious hours of German, Kafka imbued us with 'alienation', the sense of enduring hopelessness that was an inescapable fact for those of a sensitive nature.

Alienation – that was the nagging awareness of the gaping chasm between the inner world and the real world beyond, the ominous feeling that your own mind would be unable to find a firm footing anywhere in this world. All meaningful links in our tarnished world were severed: those between human beings and their surroundings, between individual and society, between humanity and God, between body and mind, between self and self-image.

It was the artist who rose in rebellion against all these impossibilities, who consciously or unconsciously revolted against the injustice called life.

Some of the artists we were introduced to at our provincial school attacked society and tried to bring about radical revolution – with violence if need be. They knew how to lend their despair and nihilism the semblance of pragmatic engagement. Since life was impossible, society had to be wholly reinvented. And in order to do that, it first had to be destroyed.

Other writers we were introduced to had a broader outlook, extending far beyond the horizons of civic critique. They realized that someone who had single-handedly severed his or her meaningful ties – with personal roots, with one's past, with God – made him- or herself a laughingstock. These writers found a kind of hysterical delight in rubbing in humanity's personal insignificance. These were the popular absurdists, such as the early Beckett and Ionesco; they knew that human consciousness was just drifting about in a sea of senselessness. They also knew that every gesture that took itself seriously was hopelessly ludicrous. It was no coincidence that most of them were playwrights: in their delirious universes everyone was an actor desperately seeking an author.

The humour of the absurdists was not without compassion. All those wandering souls who believed that their thoughts and actions were still meaningful in this world, who did not realize that they had become the plaything of forces over which they no longer had the least control, who did not want to admit that the individual had become a blank shell that could no longer accomplish anything whatsoever in the world, since that person had long ago been enslaved by the totalitarian mechanisms into which he or she had personally breathed life and over which he or she had no control: it would all be hilarious, were it not so desperate.

So there was still something bearable in the despair of the absurdists. They realized how comical it was that a human being actually had every reason to commit suicide and yet usually did not. *Waiting for Godot*: that sat surprisingly well with me, a schoolboy in Badhoevedorp. I was keen to believe that, in this senseless life, alienation was meant to be my natural state. Like so many of my contemporaries, I did not really feel at home in the world as it was. What went on in my mind did not sit comfortably with what I seemed to be in everyday reality. The reality stretched out in front of me like an endless plain, something like the nondescript polder landscape I cycled through every morning and afternoon. That estrangement learned at school was – odd as it may sound – a pleasant feeling. The writers I was expected to read and the art set in front of us taught us – while it was certainly true that we were suffering from profound alienation as individuals – that all the others, society at large, were blissfully aware of nothing. Those others had resigned themselves to the oppressive mechanisms of a society whose sole objective was the destruction of the individual.

So you could feel fairly special, a bit like one of the chosen few, in your affected despair – and you could regard others as unknowing and thus vaguely threatening victims. With all that alienation, society was something I could not belong to – and of course I had not the slightest desire to belong to it.

There were also more extreme cases in art. They were the artists who, either voluntarily or because they felt forced, had completely withdrawn into their own heads and subsequently imploded: the nutcases, the psychiatric patients, the schizophrenics and, of course, the suicide cases. These were all the rage in the 1970s. If an artist was not personally disturbed, then he made art *about* the disturbed. He wrote plays about the Marquis de Sade in the madhouse, or wrote a novel about a psychiatric patient. In German we studied Heinar Kipphardt's then highly popular novel, *März*, which – from what I remember – consisted of the poetic observations of a mentally disturbed individual who was closely studied by a doctor. The gist was that the subject's poems and drawings could tell us more about ourselves than all newspapers and news broadcasts and political statements put together. März was not a madman; he was a visionary. His terrible schizophrenia created an aperture in his mind, through which this tortured soul was able to escape impossible reality. At the end of the novel he was dead – I think he set himself ablaze. That was tragic, but at the same time it represented liberation, since for März – as for us – the world was indeed intolerable. The only appropriate end for an artist was a self-determined one. And should that not succeed, then a tragic accident that, once again, thickly underscored the senselessness of earthly existence.

That all that literary vehemence had, in reality, little to do with our day-to-day existence, that all things considered we, without exception, led pretty humdrum lives, in which we carefully planned our untroubled future without concerning ourselves too much about the impossibility of life – I wonder whether we took that into account for even an instant. We were obviously already practiced in the mindset to which the Dutch writer Frans Kellendonk would give a name only a decade later: 'sincere dissimulation'.

How did my alienation evolve? Via student theatre, a couple of years later, I inevitably came into contact with Antonin Artaud and his 'Theatre of Cruelty'. In those days, members of student drama clubs studied Bataille, Alfred Jarry, Peter Handke, Peter Weiss and Artaud. I read the vehement essay that Artaud published in 1947, a couple of months before his death: *Van Gogh, le suicidé de la société*. In this poetic diatribe, the mentally unbalanced Artaud identifies himself with Vincent van Gogh as a genius – a victim of the society that had declared him insane and pushed him to commit suicide. The big culprit was Dr Gachet, the manipulative medicine man of the village of Auvers-sur-Oise, to whom Van Gogh had appealed for help a few months before killing himself. The doctor had wanted to quash everything that made Van Gogh exceptional and had driven him straight into the arms of death. Van Gogh turned out to be an elder brother of März. The chaotic insight and liberating visions of the lonely painter who had turned away from the accepted conventions of bourgeois art, the sole objective of which was control and repression, were beyond the pale for society. That is why Van Gogh had to die. Thirty years after its publication, the poetic essay by Artaud was once again ideally suited to the wholly derivative *Zeitgeist* of my youth: Van Gogh was the exceptional loner who was broken by a society intent on quashing the individual or pushing one toward death.

In retrospect, it is not difficult to explain this fascination with the artist as a crazed genius. I belonged to the generation that had missed the last ideological boat. My predecessors had still had faith in a better world; they had been able to throw themselves with heart and mind into democratization processes; they had thought they could permanently broaden their minds by means of Oriental philosophy and a handful of stimulating substances. They had, in short, been able to see reality itself as the land of unlimited possibilities. They cherished the belief that this reality could actually be transformed. That euphoria had only just dissipated when I appeared on the scene.

In the Amsterdam of the early 1980s I was surrounded by contemporaries who spray-painted 'NO FUTURE' on walls. On my way to lectures, my eye repeatedly caught the aggressive logo of Dr Rat, a secretive presence who was omnipresent in the city in those days, and who turned out to be the foil for a young, extremely innocent-looking punk. One day he was found dead in a squat. He did not leave behind hundreds of paintings like Van Gogh, just the name of his alter ego on hundreds of walls in Amsterdam. The squatter movement he belonged to may well have championed wonderful ideals, but the hours I spent in the oppressive bastion of 'The Great Emperor', the city's biggest squat, taught me that it was primarily underpinned by a claustrophobic nihilism. The only way I could interpret the riots during the coronation of Queen Beatrix was as an eruption of a desperate urge for destruction. The chasm between word and deed had become unbridgeable; the purportedly idealistic arguments were a cover for pitch-black discontent. All around me I saw contemporaries who believed in absolutely nothing, no longer even in art. In one fell swoop, the debunking of myths that was already imprinted in me during my schooldays no longer seemed like a comforting fantasy with which we could brighten up our treacherously uniform lives. The self-destructive tendencies that had still seemed grandiose and compelling during my school years looked suffocatingly pale and pathetic in reality.

You could, it seemed, also easily go insane without being a genius. The human mind is infinitely adaptable. Those same contemporaries who, in their youth, had cherished the 'No Future' ideology as the only enduringly valid mantra later turned out not to have the slightest qualm about limiting their outlook to the here and now. In the years that followed, they pupated into materialists *pur sang*, and from surprising newspaper articles of later date about their about-face and subsequently successful careers, you could conclude that they primarily made their living in the IT sector, the business that brings the future into the present like no other.

And as for art? The antithesis of Romantic genius and oppressive society soon faded. With the unexpected reassessment of materialism in the 1980s, it seemed that the tension between personal conscience and the outside world had dissolved. Vincent van Gogh was no longer the lonely artistic talent who was cast out and fled to the embrace of insanity, but the champion in the 'booming art market'. In 1987, his painting *Irises* (1889) was sold to a Japanese collector for almost 54 million dollars, and his portrait of the ogre described in Artaud's essay, Dr Gachet, entered the annals of history as the most expensive painting of all time. The name Van Gogh no longer evoked a world of artistic visions and liberating insanity; his work was now best encapsulated by the dollar sign.

An artist who is symbolic of society's attitude toward art during the late 1980s and early '90s is Jeff Koons. His art was primarily an art of reconciliation, squaring up all the contradictions that were still desperately held high during my school years.

The world was fine as it was; there was no longer any conflict whatsoever between the mind of an individual and the outside world. Koons's glistening puppies, his childish pink pigs and gilded Michael Jackson sculptures were the final curtain for the Western Romantic tradition of art as social critique. In the world according to Koons, the individual had absolutely no desire to be unique, or for personal expression, but longed to become simply and solely a surface. In this delicious new world, art was no longer in conflict with mass culture; it had become a permanent part of it. The publicity and controversies around Koons's public persona and his art usually involved money and scandal. His critics slated him as a slick conman who – with the inflated conceitedness of the art world – was pulling people's legs by declaring that the emperor's new clothes were art. That does him an injustice; in my opinion, Koons is a brilliant artist who, in the wake of Andy Warhol, exploded the now untenable myths with which the avant-garde had perpetuated itself. These were the same Romantic notions about an intolerable life and art as an anti-societal force that I had been fed at school.

Everything that modern art felt threatened by – the flattened emotion of kitsch, the fleeting sensation of pornography, the banal uniformity of popular culture – was openly embraced by Koons, sporting a beatific smile on his lips. What was so wrong with superficiality, if it made you happy? Society was no longer a menacing monster that strove to destroy everything that is distinctive. It was, on the contrary, a world plush with soft surfaces, to which individuals simply had to abandon themselves in order to drift around forever in, nay, not a sea of senselessness, but an earthly Nirvana of reassuring superficies. All conflicts were resolved. And people were not at all unique; they were all the same, which was what connected them with each other. Why, for heaven's sake, would you want to hold on to something as inconvenient as personal conscience? Modern mass culture was the great leveller. You had to be crazy or perverse to fight it.

To this day, it remains difficult to gauge to what extent Koons was a sublime and ironic dissembler and to what extent his naive joviality was play-acting. What he was certainly serious about, and where in my opinion he was also correct, was his notion that in humanity there lies a deeply rooted desire for vacuity, nothingness, the desire to be *nobody*. And for someone like Koons, there was no essential difference between the blissful unconsciousness of Nirvana and the commodious superficialities of materialistic mass culture. The material, which modernist art had rebelled against so fervently, was rendered equal to the immaterial: humanity was equal to matter.

With artists like Warhol and Koons came the realization that art stands powerless in the world. The Romantic notion that art could truly intervene in the life of society and would be able to effect a transformation – which, with a genius such as Richard Wagner, had flourished to become an article of faith during the nineteenth century – was smashed to pieces against the wall of unyielding reality. History seemed to care precious little about art, and not a jot about human nature. The engagement that had driven so many artists into the arms of politics and activism until way into the second half of the twentieth century had drawn to a close with the messianic ideologies which had held that century in its grip – and has since then been a source of inexhaustible nostalgia. The notion that art had something to achieve in the world had turned out to be ineradicable, but what exactly was that notion again?

With the demise of Romantic engagement, the importance of the artist as a hero of society also waned. Art had not made good its promise as a saviour. Instead of the artist as visionary or as society's conscience, for the modern-day artist there seems to be but a single role remaining: fashionable fad. The artist no longer counts as the unrecognized symbol of resistance against the world as it is, but, if fortunate, the artist is adored by that same world. The work of the popular artist is permeated by the preoccupations of the *Zeitgeist*, even if critical in tone. That was always the case, but what seems relatively new to me is that society has shown mercy to the Romantic contrariness of the artist, capriciously either applauding it or summarily ignoring it. In the same way as advertising has appropriated all the big words in our world, words that had previously seemed reserved for literature or philosophy, the catchphrases of the artistic avant-garde have been absorbed by the culture of the masses. That rapid acceptance seems attractive, but even the most contrary art is ultimately a product in a mass culture, and nowadays most newly discovered art endures as briefly as the newest 'generation' of television sets.

An example: the computer brand Apple places advertisements with photographic portraits of patent twentieth-century geniuses, such as Pablo Picasso, Albert Einstein and Maria Callas, and wonders where we would be if they had not been a bit crazy. The message – hardly subtle – is that the company and its latest generation of laptops are also examples of slightly crazed genius, that they have acquired something of the charisma of all those figures of greatness, but only by grace of the inventiveness of an advertising professional. That brutal and unquestioning appropriation of the authentic by what cannot itself be authentic or unique, since it ultimately pursues commercial ends, is called 'lifestyle'. Lifestyle is materialism disguised as art. It is by far the most important Western ideology of the twenty-first century.

At the same time, postmodern art is left dumbfounded. It has been deprived of the Romantic illusion that it can really change the world. The avant-garde is tilting at windmills. The modern-day fashionability of art, egged on from all sides by society, is constantly reminded of its wretched meaninglessness by that same society. If art is actually no different from entertainment, then why should entertainment not be allowed to be called art? And what exactly is the difference between a photograph in a museum and an artsy advertisement for Prada shoes? In the eyes of society, art itself is constantly underscoring its self-evident importance, but no answer is forthcoming when one asks why that should be the case.

In the Netherlands it is remarkable that ambassadors for the arts are so poorly equipped to present a defence when art itself comes under fire via the backdoor of politically driven arts policy – and it makes no difference whether that attack is launched by right-wing populists or left-leaning councillors. When it was pointed out to him that there is a complete absence of the word 'art' in his government's manifesto, the Prime Minister of the Netherlands was able to declare, without eliciting protest, that all people have to do is look at the Erasmus Bridge in Rotterdam, since it is also beautiful. 'Art is important' is what the counter-plea of the art world always comes down to, though the whys and wherefores are practically never explained. To me it seems that the specific task of socially conscious representatives of the arts is to be able to explain to the uninitiated what the significance of art is, just as an elected representative of the people should be able to explain why a civilized country ought to reject the death penalty when a populist stands up to cast doubt on that principle.

The root cause of that half-hearted defence of the arts, I believe, lies in the fact that people within the art world are now barely capable of formulating an idea about what art is, and most especially about what art might be. You could argue that art pulled the carpet out from under its own feet with modernism. In the visual arts there is an unrelenting debate about what can actually be called art these days, now that the walls of the gallery or the museum seem to provide no guarantee of the status of a work of art.

Nowadays, what is or is not considered art is purely and simply decided by context, argues the American art philosopher Arthur C. Danto. He came up with this notion after a long, soul-searching struggle with Warhol's renowned Brillo boxes, which acquired a different meaning, when exhibited in the art gallery, from when they were stacked on the shelves of the supermarket. However, over time the term 'context' has proven to be dangerously elastic, and with so-called 'conceptual art' the visual arts seemed to have hauled in the Trojan Horse. Everything can be attributed a meaning, providing one searches for it long enough. And therefore art can also be made from anything and everything. That 'anything goes' has spawned a malaise that persists to this very day. People are assiduously searching for solutions: how can art as such re-establish a meaningful relationship with the world, so that it can once again be regarded as relevant to our culture?

Many artists, critics and policy makers follow in the footsteps of Jeff Koons. It is impossible to compete with the omnipresence of mass culture – and why would you? – so it is essential to embrace it as gracefully as possible. Applied arts like industrial design, fashion and – in the vanguard – advertising literally touch the world; they are not art forms that have nothing to do with society and hide away in the safe enclave of the art scene. They are, in short, art in action. And that such art is not associated with the pretension of wanting to change the world is simply interpreted to its advantage. This art wants to dress up the world beautifully, and that is already quite something. In the general confusion about the role that art might play, the doors of museums are opened wide to this applied art – or perhaps I can better term it 'worldly art'– which literally brings beauty to the people.

It is precisely the same in literature: in recent years there has been a growing resistance to so-called difficult – shall we say subsidized – literature, which is actually about nothing, just like so much pretentious visual art. So there are also more and more writers who place the amusement that their books offer the reader as the first priority. Since when are you not allowed to simply enjoy a nice book? The touchstone of that enjoyment is the thriller, with a beginning and preferably an unexpected end. This genre of literature is the equivalent of applied art. It is art that wants to be present in the world, but without making its mark on it.

At the same time there are the artists who think that art does not have to bear any relation to reality whatsoever, that art in itself is enough. A while ago, the poet Ilja Leonard Pfeijffer, editor of the Dutch literary journal *De Revisor*, argued that art is not obliged to be or do anything and made a plea for art that wants to be nothing but artificial. Don't inquire into the usefulness of art, he advises, for it does not have any. His literature does *not* strive 'to unburden, give an account, convince'; all it wants is 'to cruise along in the ocean of other literature that is its element and be astounded by the virtuosity of its performance.'

The critic Arnold Heumakers called this statement nothing but a reformulation of the age-old principle of *l'art pour l'art*. Art is sufficient in its own right; it is its own justification. The only thing that counts, in the words of the twentieth-century art critic Clive Bell, is 'significant form', the aesthetic satisfaction that derives from the work of art.

That much is true: art does not have to be or do anything. Above all, art must allow nothing to be dictated to it by society. But perhaps art itself wants something as well? Heumakers seems to see the position of Pfeijffer and his kindred spirits as a slap in the face of the all-powerful mass culture, an aristocratic dismissal of the vulgar questions about utility with which art is repeatedly confronted. That seems to me like a misapprehension, since mass culture primarily reveals a leaning towards beauty; it is aesthetic to the very core of its being. The popularity of design and advertising and the Western cult of fashion sprang from a desire for meaningless beauty that, in my opinion, does not essentially differ from that of Pfeijffer, except that his chosen forms are fed by a literary tradition that no longer seems to play a role in mass culture. Mass culture has forsworn literary culture and has become a convert to image and commerce, but even so I see no intrinsic difference between the playful objectives of Pfeijffer and those of Disneyland Paris.

Of course there is a difference, Pfeijffer would insist. His art is an individual art, which, moreover, requires the active involvement of the people who seek it out because of its complexity. That is certainly true, but what his views about art have in common with Disneyland – fortunately the poetry that Pfeijffer pens is a completely different story – is that they both emphatically turn away from the tangible world. It is my belief that an artist like Jeff Koons is the natural heir to the aesthetes of the *fin de siècle* – their artistic ecstasy takes on other forms, but the substance is the same. In that respect, aestheticism and mass culture go hand in hand, even though every sincere aesthete will abhor that notion.

The writer who exposed the flip side of that contemporary, materialist-aesthetic world view is the controversial American author Bret Easton Ellis. In *American Psycho*, his only masterpiece, he depicts a society, New York in the late 1980s, that has wholly converted to a life on the surface, as Koons did. Bateman leads a worldly life *par excellence*, but he no longer *experiences* anything. Yet where Koons the artist does not see anything amiss, the materialistic perfection of financier Patrick Bateman brings forth monsters. His mind proves to be a bloody cesspit where all the emotions and instincts that have been ironed flat in his world of appearances reveal themselves with an irrepressible violence and cruelty. In his imagin-ation or beyond – he personally no longer makes any distinction – he begets bloody orgies, ever more perverse, ever more extreme. The emptiness within him proves to be a deception, and all manner of primitive urges assert themselves with a vengeance because the reality surrounding him has been rendered utterly superficial.

Everything around Bateman is beautiful and expensive, and is also wholly universalized by the

laws of mass culture. He is the empty shell that
Koons wanted to be, but Ellis, who is initially seen
as a product of that same mass culture because of
his grotesque interest in its expressions, demonstrates that life based on appearances is tied
up with an urge for destruction, in which any
distinction made between fantasy and reality no
longer pertains. In his world of the materialistic
aesthetic, any true contact with reality is impossible. Ellis, unlike Koons, is a moralist.

It has often been argued that society is eager
for more and more reality in art itself: the true
story, the personal account or outpouring, the
familiar realism, the journalistic themes that
are easily transferred from the opinion pages of
newspapers to literature. Contact with reality is
being sought ever more frequently and ever more
emphatically through art. Paradoxically enough,
this contact is no longer to be found in everyday life.

It is this one-on-one recognition against
which Frans Kellendonk took a stand in his 1986
essay 'Idols', in which he also posited his now
famous but also often misunderstood philosophy
of 'sincere dissimulation'. Art is by definition
artificial, he argued, and when it tries on the factu-
ality of journalism for size, it over-simplifies the
elusive reality in which we exist. 'My objection

to realism is the same objection as that of Jews to iconic images: the pretence of knowing what we are talking about disparages the mystery,' he wrote. 'Idolatry is the unwillingness to take up the challenge of the mystery.'

Kellendonk's essay is often interpreted as a defence of the stance of *l'art pour l'art*, and it is true that here and there in his essay he seems to wallow in the realization of the artificiality of all art. However, he allows no misunderstanding about his dismissal of extreme aestheticism as well as journalistic realism: *both* fail to do reality justice.

'Just look. Here we are suddenly standing at a meeting point of two extremes. The novel that acknowledges no reality outside language, which thinks that everything exists in order to result in a book – a book that, for want of readers, must remain unopened for all eternity – and the realism that, under the guise of love of reality, attempts to bring reality to its end. Both deny the secret, both strive for an inert state of art that has swallowed reality, a satiated art for the sake of art.'

There was something of Van Gogh in Frans Kellendonk, though not the Van Gogh I had become acquainted with during my student years from Artaud, but the true Van Gogh, who perceived reality as a mystery, just like Kellendonk. Van Gogh wanted to approximate that mystery in his art without simplifying it. Neither Kellendonk nor Van Gogh denied the existence of that reality, the reality in which we are born, love and die, but they both regarded it with awe, as something that was ultimately unfathomable. All representations of that reality, whether in a painting or a novel, were tempered with reservations. That is why, according to Kellendonk, a work of art always has to display its awareness of its own artificiality as well, show that it is not reality itself. Look at paintings by Van Gogh and you can see that this is precisely how he thought about it.

What Kellendonk does not explore extensively in 'Idols' is the aestheticization of reality itself, perhaps because this had not revealed itself fully by the mid-1980s. He makes a telling comment about the difference between artistic emotions and true emotions: 'In a film you are surrounded in a determined, finite world. When your heart is set pounding in a cinema you are powerless. That is why you want to get out of there and you become agitated. Out on the street you are free to offer first aid and call an ambulance. You would only have attained idolatry if you were to turn away your head out in the streets as well.'

That last, irony-drenched sentence is significant. For the main character in *American Psycho*, and indeed for all of us, reality itself has increasingly become a determinate world, as if we were actors in a film. That is also what you always hear when something unspeakable has happened in the real world, when passers-by have stood there, looking on dumbfounded, while someone was kicked to death in the street or when two aeroplanes smashed into the World Trade Center in New York: 'It was just like a film.' That last event *is* indeed a film for us and, moreover, it has been repeated *ad infinitum*. We have the feeling that we have witnessed something horrific and yet we stand there and simply look on.

It is that unreal reality from which art should shake us; not by striving for realism, which is already ubiquitous in our image and information culture, but by reconnecting our numbed consciousness with reality as a mystery.

Postmodernism in art has long been seen as a perverted form of modernism, a conception of art in which art history serves as an enormous warehouse where everything is ultimately all the same – a vacuous game devoid of inspiration. The totalitarian principles of modernism and its leaning towards absolute abstraction may have been abandoned, but by no means renounced. On the contrary, what predominated was the sense that modernism had achieved something that it would be impossible for artists to match thereafter – somewhat similar to the way we schoolchildren of the 1970s were made to believe that we would never be able to surpass the 1960s.

I think that the time has come to redefine postmodernism and learn to see it as a reaction to modernism, a reaction that is radical. Postmodernism specifically seeks out the reality to which abstract modernism had haughtily turned its back, denying even its existence. It is a reality that mass culture essentially wants nothing to do with either.

My postmodernism very much recognizes the existence of that reality, intangible though it is, and no longer renounces it. My postmodernism sees it as art's mission to be passionately engaged with reality, to see the world as new. That realism belittles reality is a lesson of modernism that we take to heart, but this does not imply that we, along with realism, must be disdainful of the concept of reality as well. On the contrary, we must approach reality like trembling unbelievers in the presence of something immeasurable. To experience that you are alive, that you are part of that awe-inspiring universe known as reality, the realization that the depths of the human soul, as Saul Bellow put it, do not exist in isolation, are not limited to our imagination, but are intimately connected with that other great unknown: the world in which we live.

With art on the brink of surrendering to the warm bath of mass culture and thus rendering itself superfluous, this seems to me a task that is not only honourable but also necessary. Such art will not change the world. What it can achieve is to open our eyes, build a bridge between what takes place inside our minds and beyond. Such art will make us forget the sensations of mass culture and restore to us the experience of the mystery of reality.

Cornel Bierens: We'll slide down

A few weeks ago this newspaper published a large colour photograph of a French woman painting her lips. The beautiful young woman was supposed to draw attention to an article about the lives of rich kids belonging to the Paris jet set, as described by young French female novelists. It was the kind of photograph you used to see only in fashion magazines but for which serious newspapers, too, now happily sacrifice column space.

As of last week, a dead ringer for this photograph is on display at De Vleeshal's picture exhibition *We'll slide down the surface of things*. The woman now looks the other way, painting her eyelashes instead of her lips, but for the rest the painting looks exactly like the photograph. The features, the inward-looking gaze, the mood and composition – the painting and the photograph are indisputably mirror images.

In a conversation with the artist, Düsseldorf's Frank Bauer, he shows no surprise at the striking resemblance between his work and the, to him, unfamiliar photograph. He just happens to take his inspiration from magazines, videos and adver-tisements, all brimming with such images, so these things can happen. 'I want to be as contemporary as possible,' he says, to him the photograph only proves that he has succeeded.

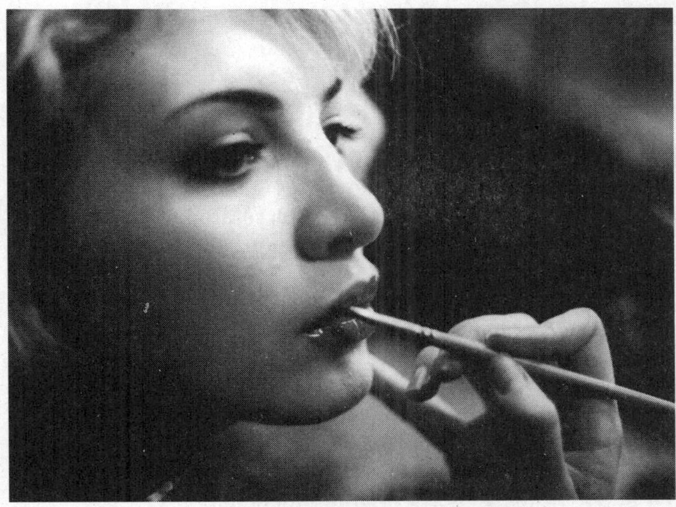

3.1   Edward Holub, *Woman Getting Lips Painted*

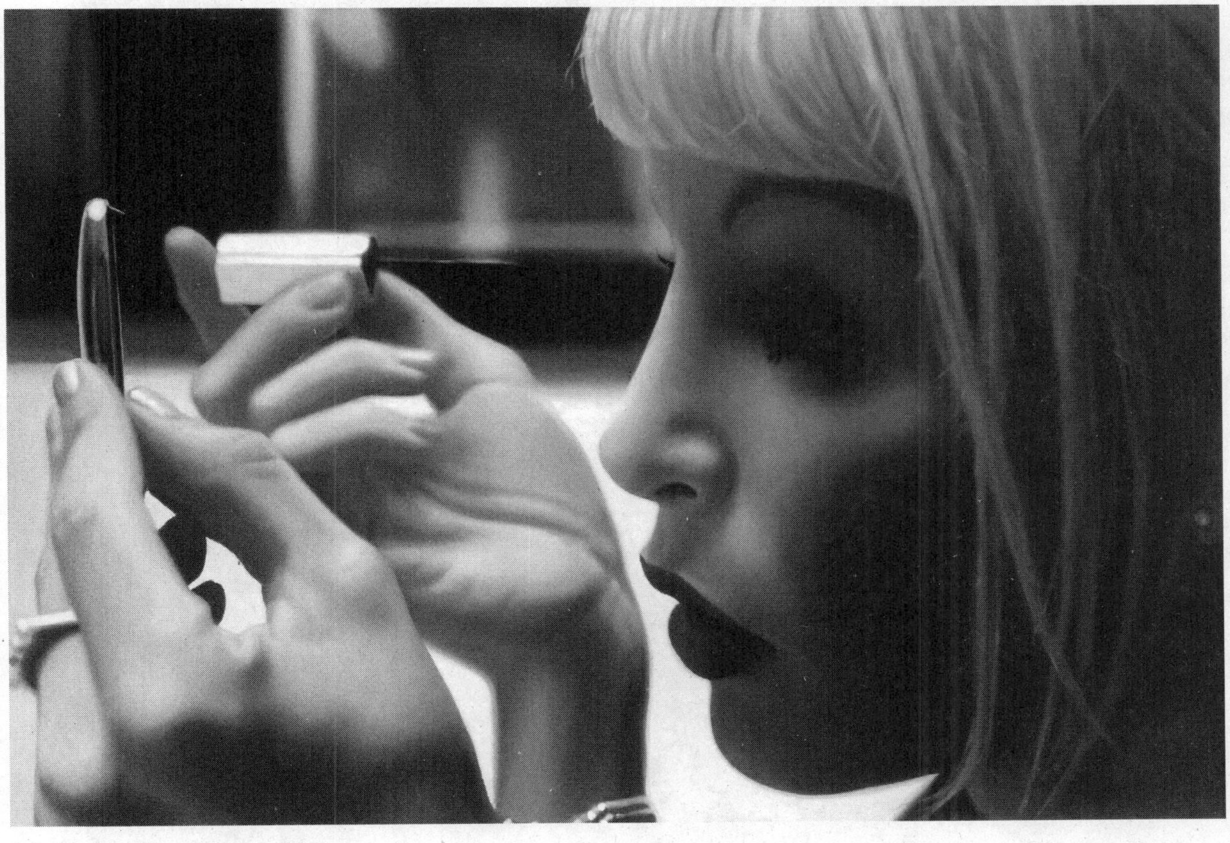

3.2   Frank Bauer, *Sasha*, 1999

Bauer (b. 1964) is a contemporary of Bret Easton Ellis, and, judging from his work, also a kindred spirit. Ellis is one of those American authors who have shown the newest generation of French women writers how to convincingly portray bored and wealthy young people. With his novel *Glamorama* he also supplied the idea and the title for the exhibition in Middelburg.

The first-person narrator of the book, Victor Ward, is completely uprooted psychologically and can only deal with his experiences if he can relate them to images, words or gestures he knows through film and television. In a limo, on his way to a party, he hears a U2 song with the line *We'll slide down the surface of things*, which continues to run through his mind all evening. It becomes some kind of incantation, providing him with a last vestige of control at the booze-, drugs- and gossip-riddled party where lines like 'I am so tired of looking at that empty expanse that's supposed to be your face' fly about.

Ellis's books can be intensely boring and yet, as cutting sketches of a time in which all substantive motivation has been cast aside, also very clever. They reveal not only the emptiness of glamour, but also the lethal forms that vacuity can take on when the desire for glamour becomes universal. If everyone longed to shine and to take

3.3 *We'll slide down the surface of things…*, installation view, 2002

centre stage, we would end up in a world where appearances ruled, simply because everything would become cosmetic.

We are living on the crest of that development. World news undiluted with women putting on lipstick has almost become inconceivable. To condemn this is pointless; it would be better to accept the predilection for appearances as a given and, in so doing, to explore whether new meanings can nevertheless be attached. *We'll slide down the surface of things* can be perceived as an attempt to do so.

In De Vleeshal architect Herman Verkerk has built a special exhibition space for *We'll slide down the surface of things*, consisting of three half-open white cubes, placed one behind the other. The modernistic little structure, a mini-version of a twentieth-century museum, appears to have been hit by a fierce gale. Besides Frank Bauer, the three rooms house two other figurative painters, Dutchman Arnout Killian and American Glen Rubsamen, the latter in between the other two.

In Frank Bauer's work, as in Bret Easton Ellis's books, party going is a favourite theme. In one painting the German portrays the party-goers themselves, in another the silent evidence: overflowing ashtrays, empty bottles and mirrors

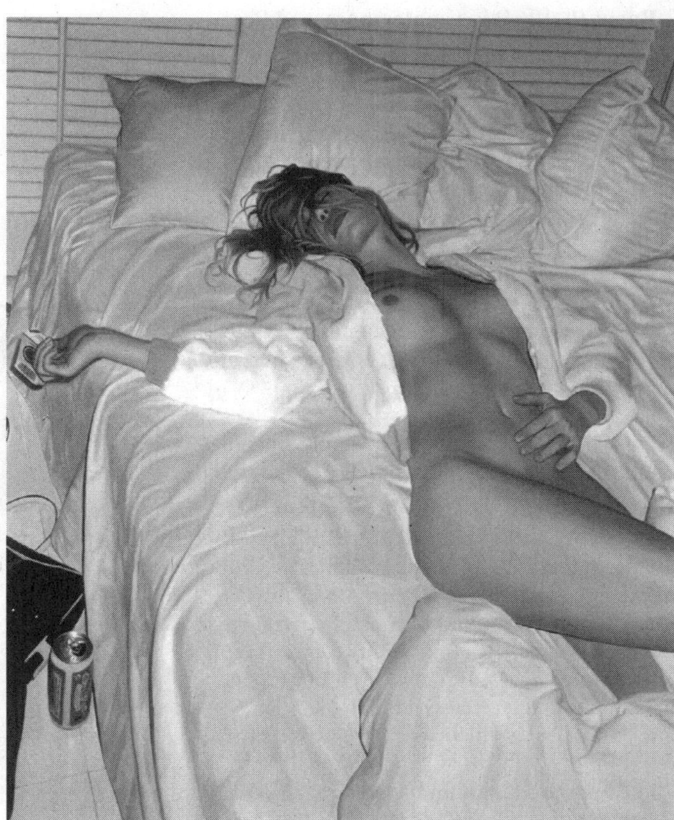

3.4 Frank Bauer, *Sasha in New York*, 2002

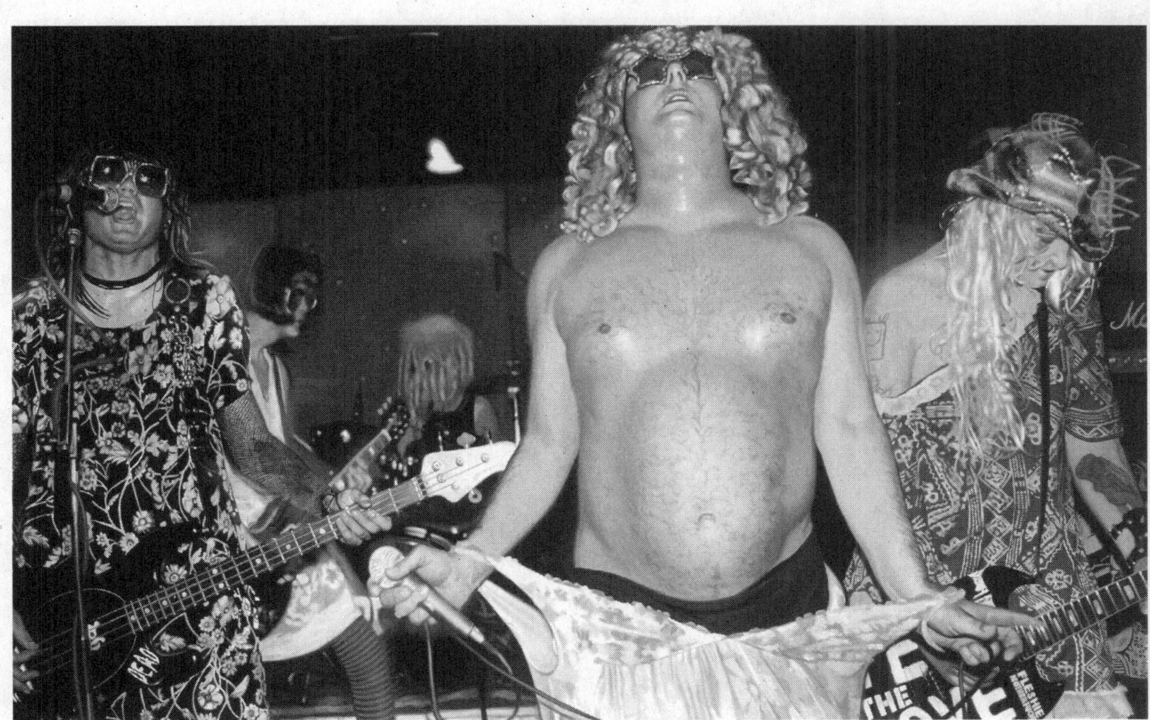

3.5 Frank Bauer, *The Grannies*, 2002

with cocaine on them, all very smoothly painted in photorealist style. A large painting by Bauer of a nude woman on a bed is also on display.

The influence of his mentor, Gerhard Richter, is marked, but when asked the obvious question – whether he has not also been formed by 1970s American photorealism – Bauer answers in the negative. He does not care much for the virtuoso painters of reflecting glass and chrome car bumpers, who he thinks 'had too little content and couldn't even paint human skin'. The latter is something he very much likes to do as he aims to make his paintings 'as personal as possible'. 'The people I portray are all friends,' he adds.

Yet the question remains what, precisely, makes his work personal, for his paintings come across as neutral, cool records in which all comment has systematically been avoided. 'I wouldn't mind commenting on decadence, waste and the like,' the painter says, 'but because I portray my own life as well as that of my friends, I can't really be critical, not even obliquely. I'd rather be detached and open at the same time, so that people can see what they want to see in my work.'

The approach is almost scientific. It seems the painter wants to convey, first and foremost,

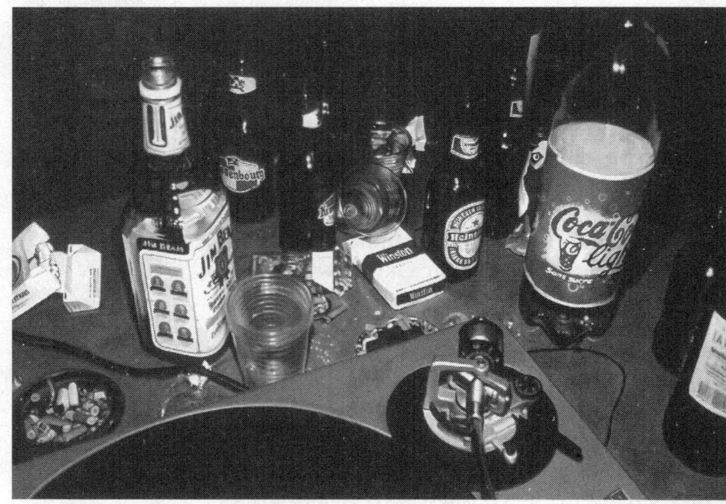

3.6   Frank Bauer, *After Hour*, 2000

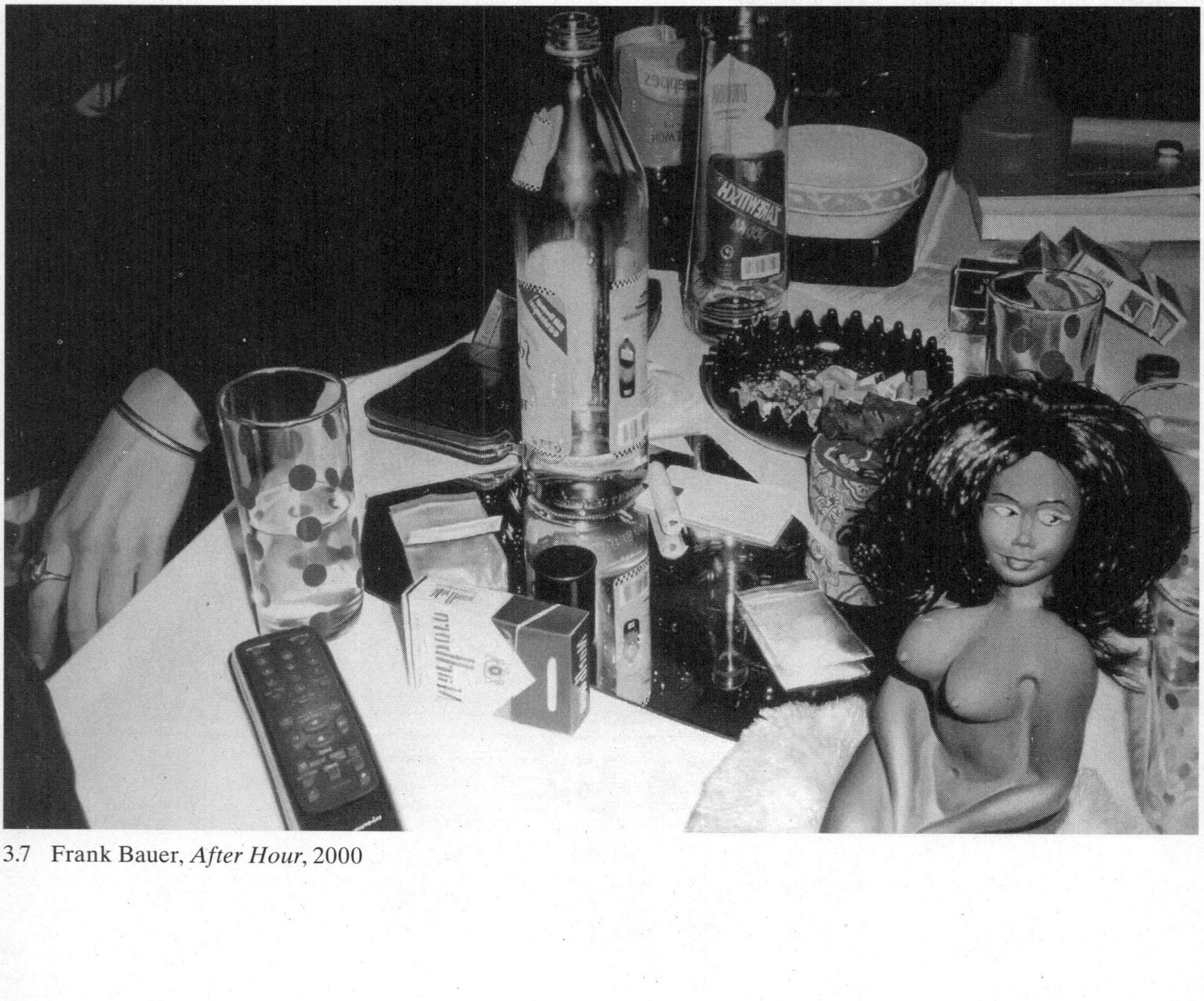

3.7   Frank Bauer, *After Hour*, 2000

that the world he portrays is simply a fact. What he wants to paint is the naked truth. But why paint? Do not all those other channels of communication present us with enough of the naked truth? Why go so far as to record all those hyper-realistic images through the cumbersome method of the brush?

Bauer's answer is short and simple. 'Because I am a painter and yet I cannot escape harsh reality. Apparently I have to work through reality in order to get somewhere else.' With this, he succinctly sums up what the exhibition is all about: reality, in all its ostentation, has become so obtrusive that painting has no other recourse but to grab the bull by the horns and face it head on. Recording superficiality straight out: that's the way *to slide down the surface of things*.

It is funny how, after one talks to Bauer, his paintings become less photographic and more like painting. All of a sudden you see imperfections, strange shadows, contours out of place, anatomical frictions. These could be indications that the painter is already, covertly on his way to that 'somewhere else' he hopes to end up. His imperfections could be small rodents secretly undermining the depicted world, eventually providing the commentary that the painter as yet fears.

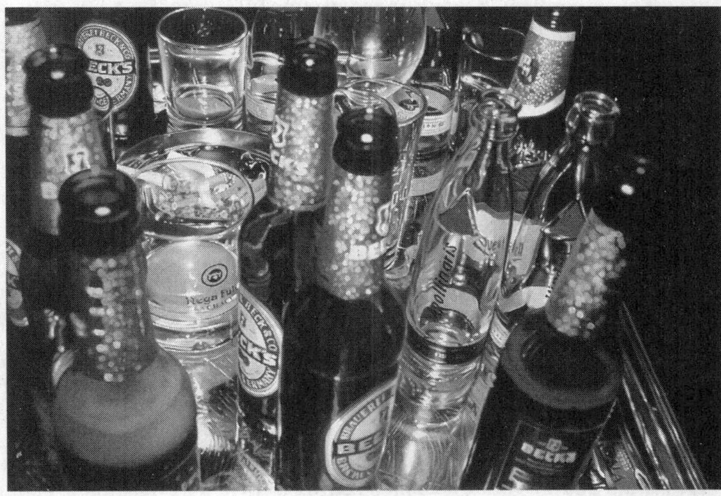

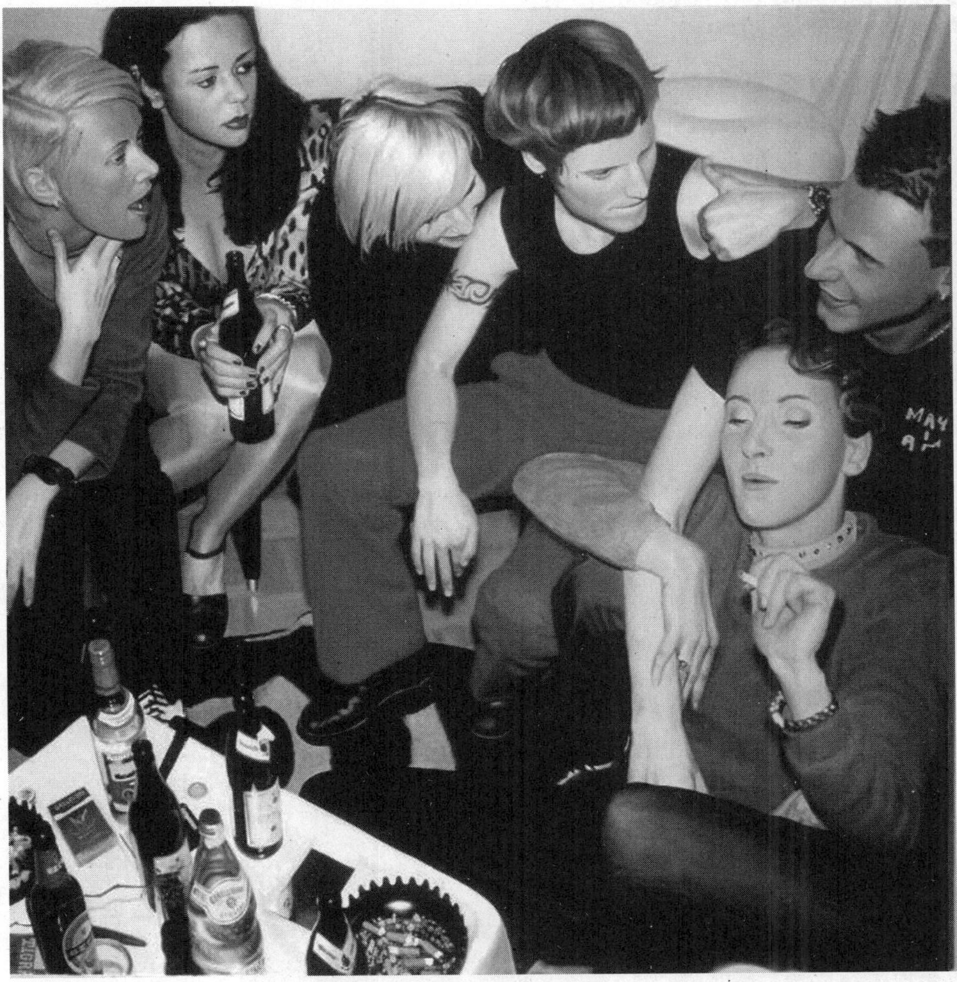

3.8   Frank Bauer, *After Hour*, 2000

3.9   Frank Bauer, *After Hour (Tussi Deluxe)*, 2000

All great art comments, therein lies its content. All beauty needs barbs in order to rise above hollow aesthetics. Painters with great technical prowess are especially familiar with the problem: they are constantly in danger of wanting to paint *beautifully, too beautifully.*

This is confirmed by an anecdote told by Arnout Killian (b. 1969), who sold a painting to Dutch interior designer Jan des Bouvries which he later saw in one of Des Bouvries's catalogues. To his dismay he found that, in that context, the work had lost something: 'It had become fashion, a prop.'

Seeing Killian's paintings, it's easy to picture the story. Like Bauer, the former Rijksacademie student produces photorealistic works, but chooses more singular topics and zooms in on them more closely. A woman's face in a face pack, a window dummy's hand, a body in a bubble bath. Professionally painted large canvasses, each and every one, but no easily discernible comments or barbs.

Yet according to Killian they are present. He points out the shadows on a painting of two women's legs walking down a catwalk, calling them 'sinister'. 'Painting,' he argues, 'is done in a primitive fashion, by hand, and so the images automatically acquire psychological overtones. That's where the commentary is. Precisely

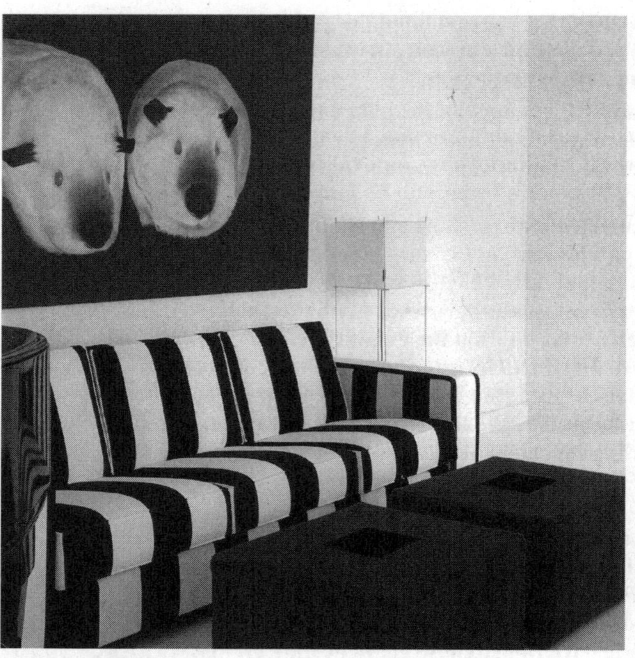

3.10 A painting by Arnout Killian used as a fashion prop and photoshopped pink for interior design book *Dutch View* by Jan des Bouvrie. *Eigen Huis & Interieur*, VNU Tijdschriften, 2001

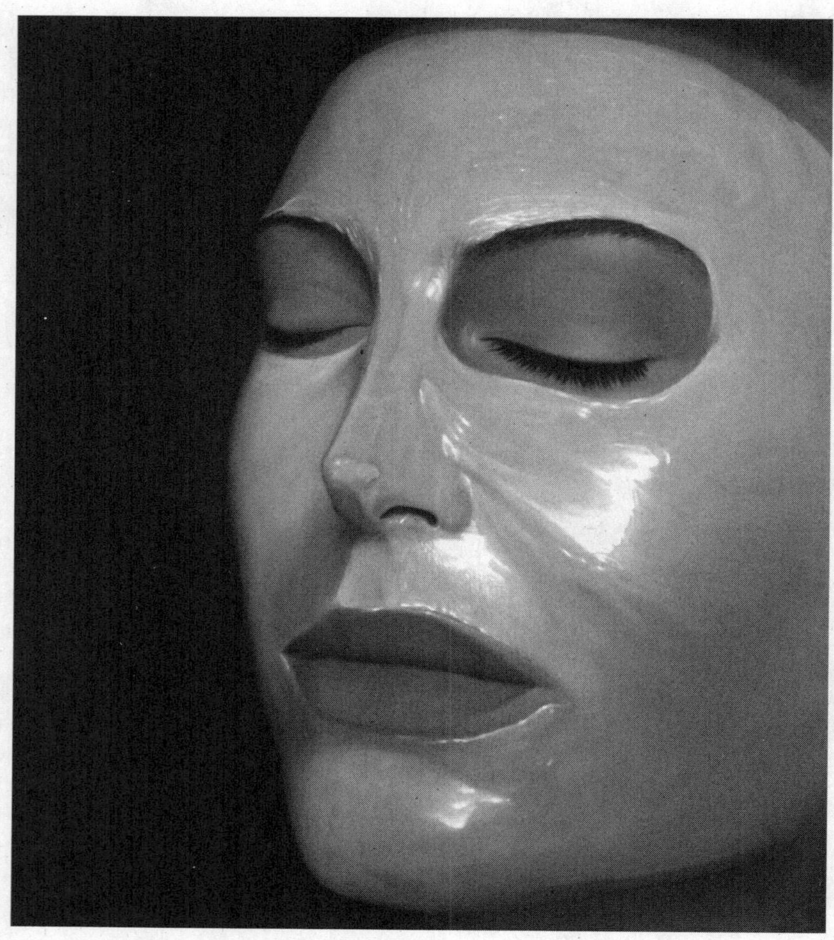

3.11 Arnout Killian, *Bleu Beauty Mask*, 2001

because painting has remained this primitive, it has the ability to comment. In other media machines always intervene.'

Be that as it may, how much can you comment if you limit your scope to sinister shadows? Shouldn't a painter be much more direct and explicit?

'You lose it if you try to be more direct,' Killian says resolutely. 'There is endless talent in the advertising and film industries, all working together. What *Gladiator* achieves is astounding; a painting could never match that. In Munch's time Munch was astounding; in that respect painting is through. But it can still provide personal comments on visual culture. Painting can freeze images, in a way that other media can't.'

On the one hand the crushing advertising and film industries, on the other the world of design that reduces art to fashion – these are the opposite poles between which the painter wishing to comment on visual culture finds himself. It is a position in which he has to feel his way between going unnoticed and being completely absorbed.

Unlike Bauer and Killian, the third painter in Middelburg, Glen Rubsamen (b. 1959), paints landscapes rather than people. His paintings are

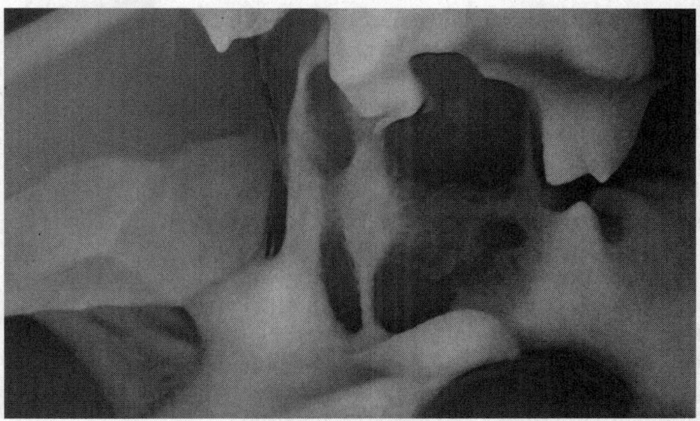

3.12 Arnout Killian, *Foam*, 2002

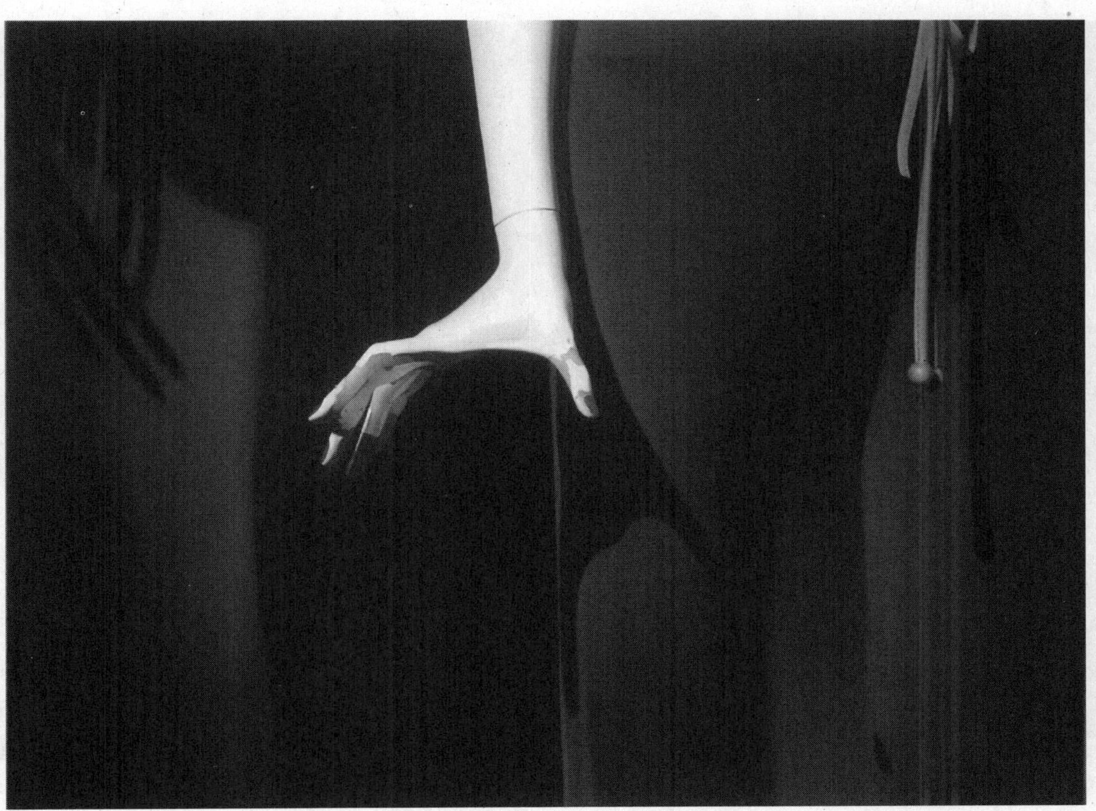

3.13 Arnout Killian, *Frozen*, 2002

modestly 'homey" in size and are influenced not by fashion but by tourist photography. Still, he feels at home with the others, much as a photograph of a palm tree does next to a photograph of a beautiful woman in a magazine. 'The first may not have as manipulative an effect and be less *pushy*,' he states, 'but both appeal to the same fantasy.'

Not as pushy – does this make trees more worthy subjects?

'To me, painting a tree is like portraying an individual. Trees have just as much personality as people. I like to look at old paintings to see how much personality has been given to the people and how little to the trees. I think painters have long been afraid of trees. Manet was the first to come to terms with them, and after him landscape painting went back into a kind of hibernation. I'm convinced that the possibilities for landscape painting are endless.'

All of Rubsamen's landscapes have been painted against the light, at dusk: the hour during which the sky is still bright and coloured and the foreground already dark. The main constituents are invariably the evening sky and, against this, a few scattered silhouettes of streetlights and other traces of human activity – and trees of

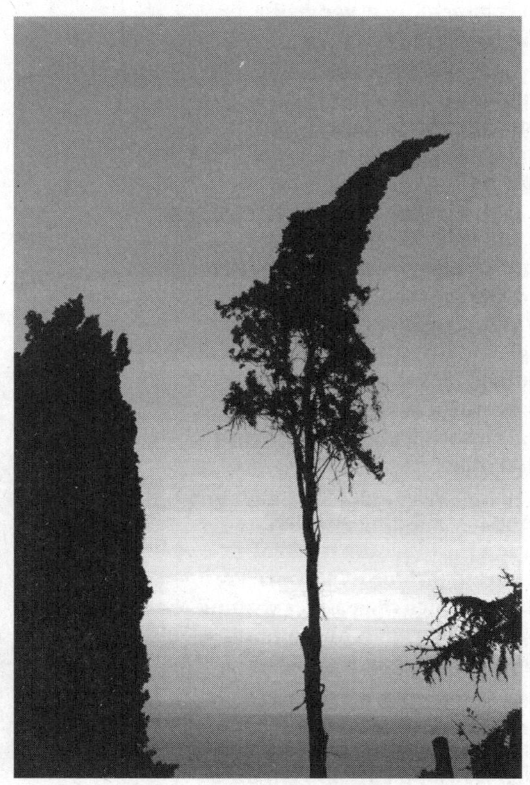

3.14 Glen Rubsamen, *The Call of Cthulhu*, 2006

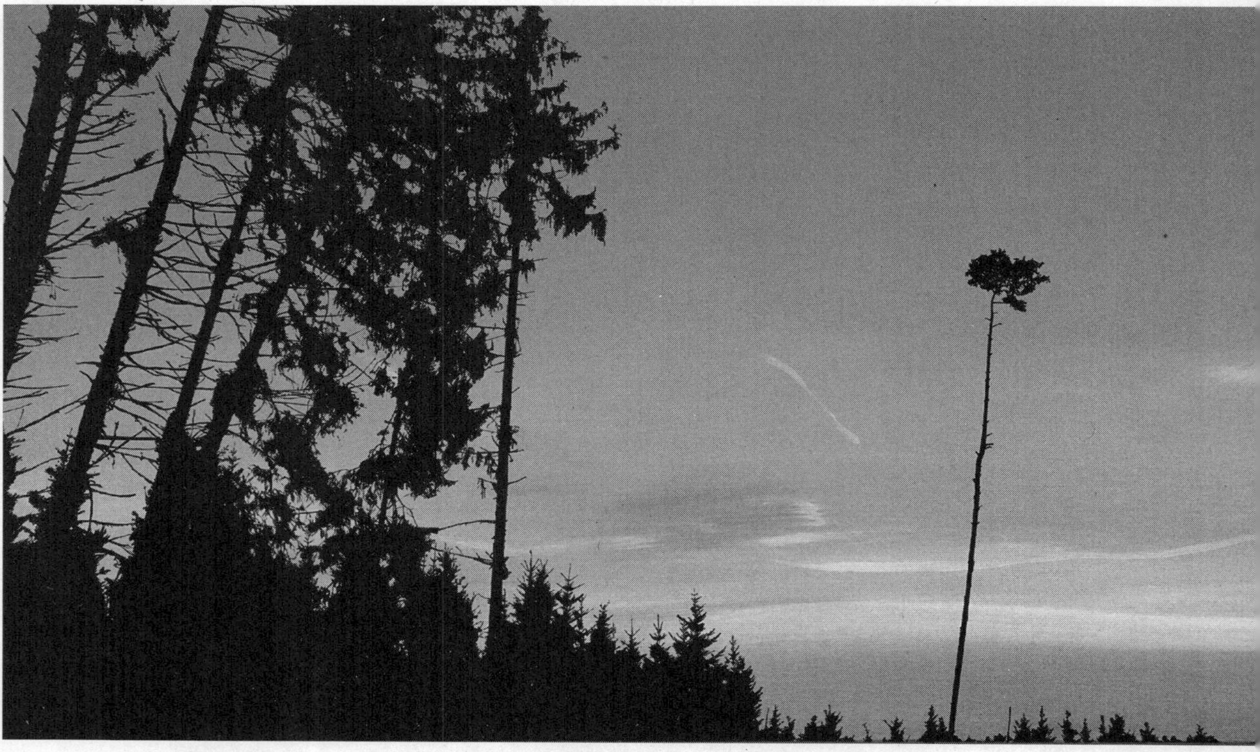

3.15 Glen Rubsamen, *Please Let My Affections Lead Me Into Danger*, 1999

course, many palm trees in the style of travel brochure covers.

It is intriguing how the American's 'atmospheric realism', as he himself calls it, succeeds in creating the impression of the paintings being both landscapes and botched holiday snaps, of the Apocalypse looking exactly like a lovely evening in the back garden.

Of the three, Rubsamen's work appears to be the most mature, for however little they show, his landscapes seem to be representative of the world, including its absent humans. Bauer and Killian, on the other hand, limit themselves mainly to the glamorous world they want to paint. Which, incidentally, is not difficult to understand, since while Rubsamen has considerably revalued his trees by giving them human characters, Bauer and Killian would only debase their beautiful women if they were suddenly to see trees in them.

And yet they will probably do so some day, once they have, as Bauer puts it, 'worked through reality'. Then Bauer will give the skins he so likes to paint the patina and knotting of tree trunks, and Killian will have branches stroll down his catwalk. And they will see that you can never go far enough in order *to slide down the surface of things*.

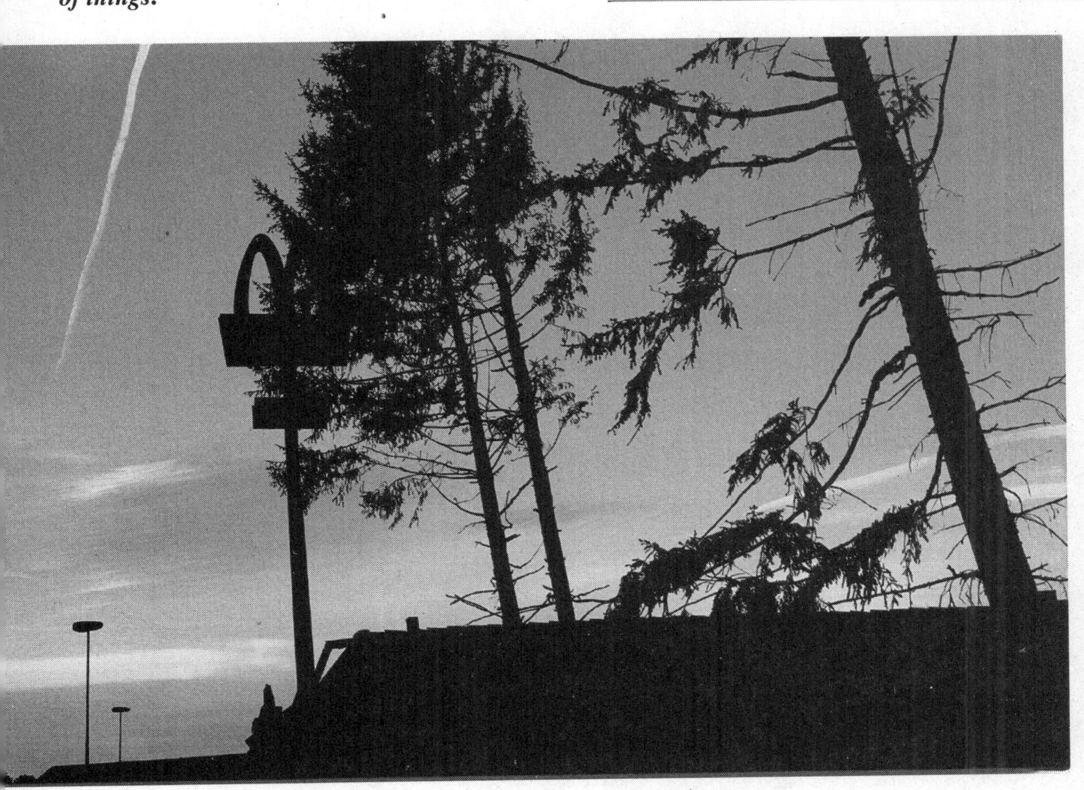

A –List for opening **'*We'll slide down the surface of things…*'**
De Vleeshal Sunday September 8, 2002, 4-6 p.m.

A		
Amis, Martin	Grossman, Martin	Prada, Miuccia
	Gursky, Andreas	
B		**R**
Ballard, James G	**H**	Ravenstein, Apollonia van
Barney, Matthew +/	Haan, Ingrid	Reijn, Halina
Bauer, Frank (Artist)	Hamer, Sandra den	Rijn, Kees
Beumer, Guus	Heer, Bregje de	Ritsema, Maaike
Blom, Ansuya	Heijne, Bas	Rolf (see Viktor)
Bouvrie, Jan des	Hirst, Damien	Rubsamen, Glen (Artist)
Brard, Patty	Houellebecq, Michel	
Brady, Ian	Humobisten (2X)	**S**
Bruggen, Koosje van		Schrofer, Jan Willem
Byvanck, Valentijn	**I**	Simons, Raf
		Slobbe, Alexander van
C	**J**	Sotos, Peter
Carels, Edwin	Jakobsen, Paskal	Spencer, Jon
Celante, Germano	Jonze, Spike	Stoner, Tim
Cijsouw, Jacqueline	James, Richard D	
Colburn, Martha (Artist)	Jenkinson, Tom	**T**
Cosar, Michael		Tavares, Ana Maria
Coppola, Sophia	**K**	Tilroe, Anna
Cunningham, Chris	Kennis, Bas	Tuymans, Luc
Cuypers, Raymond	Killian, Arnout (Artist)	~~Trommel, Annelies~~
	Kmetic, Sasha	
	Knoxville, Johnny	
D	Koolhaas, Rem	**U**
David, Catherine	Kooten, Kim van	
Derwig, Jacob	Kristel, Sylvia	**V**
Dijkstra, Rineke		Verkerk, Herman (Architect)
~~Dibbets, Jan~~	**L**	Verhoeven, Paul
Dumas, Marlene	Lange, Connie de	Viktor (see Rolf)
	Lieshout, Joep van	Voss, Rüdiger
	Lucas, Sarah	
E		
Ellis, Bret Easton	**M**	**W**
	McBride, Rita	Ward, Victor
F		Westeinde, Adri van 't
Field, Simon		Wijker, Kees
Fruciante, John	**N**	Wolfson, Rutger (Curator)
G	**O**	**X**
Gelder, Wim van	Oliver, Jamie	**Y**
Gelder, Mineke van	Osborn, Jack	**Z**
Gladstone, Barbara		
Gordon, Douglas	**P**	
~~Grunberg, Arnon~~	Pallenberg, Anita	

3.16 Invitation card for opening, *We'll slide down the surface of things*, 2002

Bas Heijne: Real Seeing

Several years ago the Stedelijk Museum in Amsterdam showed an installation by visual artist Germaine Kruip, entitled *Two Seconds*. The whole thing was so simple it was almost invisible. A short, broad set of steps led up to one of the museum's windows; any visitor who took the trouble to climb them simply looked out through one of the building's large windows. He or she saw nothing more than what was taking place outside the museum every day: the bustling street life of Amsterdam. The installation also incorporated sound: one could hear the sounds of the outside through a speaker, but with a two-second delay, which produced a remarkably alienating effect. That was all.

What Germaine Kruip (b. 1970) had managed to produce with a gesture as simple as it was inventive was to turn everyday reality into a museum piece. Precisely because the museum visitor who got up in front of the window saw nothing special, and certainly no art object, he or she felt compelled to take another look at what at first glance seemed all-too-familiar: trams, cyclists, the hubbub of the city, passers-by, a grey or blue Dutch sky, depending on the weather conditions. One looked at reality as one looks at art.

4.1   Germaine Kruip *2 Seconds*, 2000

The effect operated in both directions, however. Through her installation, which actually barely aimed to qualify as such, Kruip also opened up the museum to the outside world. Through a simple intervention, the aesthetic domain that, by definition, a museum is suddenly engaged in a vital relationship with transmuted reality, so that one not only saw life on the street in a different way, but the museum itself as well. For the unsuspecting passers-by in the street there was also an un-expected apparition: they suddenly found themselves being looked at inquisitively or intensely by a living figure that would appear in the window of the museum. Art and life: suddenly it was no longer clear which was which. Kruip had not created art in the public space; she had created art *out of* the public space.

The intervention she performed in *Two Seconds* was minimal, and the detractors of purely conceptual art are wont to deny it the status of a work of art, as it barely seems to surpass the quality of a clever *trouvaille*. Yet it is precisely the absence of a tangible object *as art* that is the premise of Germaine Kruip's conception of art. She aims to create art that is invisible, or barely visible, that intervenes as unobtrusively as possible in the life of the spectator. 'You create reality as you look at it. I want to make my spectator aware of that,' she says of her own work. Above all, the spectator should not be aware that he or she is looking at a work of art. Small, apparently spontaneous revelations that take place in everyday life: on another occasion Germaine Kruip reproduced a falling star using a fireworks rocket, to inspire night-time visitors at an art project to make a spontaneous wish; to her it was successful only if the people who made a wish had not realized that it was something artificial, that it was *art*. In her work and in her statements, she fights against the idea of art as an autonomous object, as something ensconced, inaccessible and autistic, in its own reality.

Yet her work – and this is what particularly appeals to me about it – is not intended as a *démasqué* of the autonomous work of art, yet another unmasking of a worn-out faith in art as a carrier of elevated meaning, or worse yet, as a status symbol in a thoroughly commercialized world. While she obviously detests the idea of art as a totem, she is not out to destroy the object or play empty games with our subconscious assumptions of what art is or should be. On the contrary, it is not difficult to also see a work like *Two Seconds* as a passionate plea on behalf of art as something that can shape the world around us. In her *Two Seconds* installation and in the rest of Kruip's work, it is indisputably imagination that shapes reality.

However, the spectator is put to active use in her art; his or her passive gaze is made creative, as it were. Her art is nothing less than a way to re-establish contact between human consciousness and the world that surrounds it.

II

'You create reality as you look at it.' If art can do anything – or to put it more emphatically, if art *must* do anything these days, it is that. There has been a great deal of intense debate in the last several years about the so-called crisis in art – and whether this crisis exists or not, the intensity of the debate actually shows that there is, at any rate, a *sense* of crisis – and all of these discussions can be traced back to a shaky or vanished faith in the social significance of art.

In the nineteenth century, such divergent artists as Wagner and Rimbaud cherished the hope that art might regenerate an old and exhausted world; through art, a genuinely new world and new humanity might be born. This hope was essentially religious, and in Wagner's case his art was certainly a means to free man of all that held him shackled to a stultifying, materialistic existence. Painters such as Van Gogh and Gauguin also strove to bring the sacred back into a tarnished world through their art; the two men wanted the same thing, but their methods were radically different. Gauguin tried to leave earthly existence behind by taking refuge in his imagination (and in the fantasy of a largely invented tropical paradise, where the ugly modern world has been miraculously made invisible), while Van Gogh aimed to convert visible reality to his gaze, as it were. In his heated arguments with his putative brother in art, he insisted that he always aimed to be realistic, that is to say, that his paintings always had to have a *real* subject. A real face, real sunflowers, a real pair of worn-out shoes: Van Gogh's imagination always had to be in direct contact with reality, which he then transformed into something new and never before seen.

This great hope in art as the radical renewer of the world was not fulfilled; you might say that ever since, art has been mostly concerned with renewing itself, endlessly. The political engagement of many artists, both extreme left and extreme right, in the twentieth century was once again based on the idea that art had an important role to play in the renewal of the world, but the emphasis was now far more on a societal, political revolution. The idea of an artistic avant-garde was closely linked to the great ideological ideas about the New Man. This great hope was not fulfilled either; most ended in death and devastation. The abyss between imagination and reality opened by the major ideologies of the twentieth century echoed with the scream of revulsion that had issued, at the very dawn of that century, from the contorted lips of a dying Mr Kurtz: *'The horror! The horror!'*

Ever since, art has been singing a different tune. Classical artistic notions about renewal and the avant-garde have survived, but in the domain of *fashion*. In that respect, we have become unmistakably postmodern, keenly aware of mechanisms and inclined to separate concepts from the reality to which they refer; we can now *act out* the convictions of the twentieth-century art innovators. Renewal in art has become purely aesthetic; any claim to renew of the world and ourselves has virtually disappeared. Admittedly, desperate attempts are still made to reanimate the political engagement of the artist, but instead of creating a new élan, such attempts, all things considered, seem mostly nostalgic. The experience of the last century has made us overly aware, and even attempts to ignore this emotionally charged legacy merely demonstrate the extent of its impact.

Faith in art in general has been undermined in different ways, and today it is particularly threatened by political populism and by mass culture, which in any case are tightly intertwined.

Populist politicians keep asking what the use of art is, especially art in the public space, which is after all foisted on citizens who in most cases did not ask for anything. Why does so much money go to art that has no meaning at all for most people? One no longer speaks of an avant-garde forging ahead of the public, but rather of a deep chasm between the two, making it seem that the avant-garde of visual art now only communicates with itself. The threatening demands of the populists in fact usually remain unanswered; the representatives of the Dutch art world have a hard time explaining to the public at large wherein lies the significance of contemporary visual art for society. Most of the time they beat an exasperated retreat into nebulous jargon. One contributing factor is that many among them harbour major doubts themselves about the *status aparte* claimed by visual art in today's mass culture. The popular culture of entertainment and advertising teems with meanings and is, furthermore, omnipresent – this makes the so-called unique work of art, which in most cases sees its presence confined to the interior of a museum or art gallery, suddenly seem anemic. The excitement the avant-garde of the past used to generate is now really found only in mass culture, with its torrents of images and hypes – and in the work of those rare artists who manage to make skilful use of the channels of this popular culture.

III

As I mentioned, the public debates and protests are never more intense than when they concern art in the public space. Social resentment often plays a role in this – after all, social resentment rears its head in *every* debate about art in the Netherlands; there are few countries in which art is so entangled with feelings of wounded self-esteem and fear of egotism in others. Something is imposed on a community from the top down, which automatically arouses a mistrust that no amount of public discussion sessions can dispel. In some instances, the work of art in the public space turns into an object of aversion and aggression – because people find it, in plain language, *ugly*. But that word often masks deeper emotions: independent of the aesthetic qualities of the object erected in the public space, the object itself arouses negative emotions, simply because it is visibly presented as art. In contrast with the past, when there was vehement resistance from the bourgeoisie against works of art because they were considered an affront to good taste, or offensive, more and more often these days people are resisting the idea of art itself, because for many people this is synonymous with expensive futility and elitist pretension. In such cases it is the object itself that seems to be a tangible motion of no-confidence toward its setting.

All too often, the message seems to be that the artwork is intended to being beauty and meaning into an environment that sorely lacks such qualities. Art is beautiful; you are ugly.

Art as a palliative: a lot of art in the public space is seen as a contemptuous indictment of the people in whose midst it is erected. It is precisely the impregnable autonomy of the object that seems to strip its setting of meaning instead of imparting it. From this standpoint, it is an pathetic inversion of what an artist such as Van Gogh was aiming for; to him, art was there for the world, not the other way round.

All of these factors explain why people have become so cautious in the last few years when it comes to art in the public space. Yet instead of getting bogged down in long legal procedures and endless series of public discussion sessions, it seems to me it would be more useful to look for the meaning that art might have for our time, for our society. Works like Germaine Kruip's *Two Seconds* and countless other artworks by contemporary artists, such as the photographs of Wolfgang Tillmans, for instance, betray a relationship with reality that in many ways is reminiscent of Van Gogh's. They no longer presume to create a genuinely new reality. The aim is simply to make spectators take another look at what seems a given, to make them aware of their creative gaze. This automatically produces something new, even if only for a moment.

This new relationship with an existing reality, which paradoxically is being concealed from us by the unremitting stream of images in the media and the visual offensive of mass culture, automatically legitimizes art. Even art in public space – it is the work of art that opens our eyes and makes us look again at an all-too-familiar reality.

Mass culture and the cult of the image have turned the spectator into a passive observer. This threatens to turn the world into nothing more than impressions; the image is impressed upon us without our own imagination being put into operation. Postmodernism begins, writes Irish cultural critic Terry Eagleton in his book *After Theory*, when we cease to obtain information about the world and, instead, absorb the world *as* information. This definition is as clever as it is significant. Eagleton very accurately identifies the paradigm shift that has taken place in our culture: precisely because of the never-ending offensive of images that confronts us with the entire world on a daily basis, reality threatens to become a kind of video rental shop. When we have absorbed

the selected tapes, we bring them back. There is
a danger that we will increasingly view reality as
mere fiction, encouraged by postmodern concep-
tions of reality as nothing more than a construct.

Yet Kruip's statement about the spectator
creating reality by looking at it aims to emphasize
something beyond than the postmodern cliché that
each person creates his or her own reality. There
is in fact a world that exists independently of our
imagination: a city, a street, a new road through
an old landscape, like the N470 motorway. This
reality exists outside of us; we look at it the way we
look at a landscape from the side window of our car
or the window of our train compartment. We see
something that can only acquire depth and meaning
through our imagination, through our creative
gaze. That is what Kruip's art and that of so many
others is striving for: creating conditions that put
our imagination into operation, that make us take
another look at what is apparently meaningless,
that re-establishes contact between our reality and
something located outside ourselves, yet to which
we know we are connected in some mysterious way:
the outside world.

This sort of art does not withdraw from the
world; it does not interpose itself ostentatiously
between us and reality, but in fact makes this world

accessible to spectators and lets them see what they have not seen before – or more accurately, what they have not seen before *that way*. Art like this does not use the public space not as meaningless scenery, which only serves as a backdrop for an aesthetic object that is complete in itself. Art like this needs the public space. It is its natural domain, just as the public space is the natural domain of human beings.

Bregtje van der Haak: Interview with Tim Stoner

Rutger Wolfson (RW): Weren't you spending some time in Marbella when you started to paint these utopian-looking holiday scenes?

Tim Stoner (TS): The idea came about after seeing an exhibition of Goya in a print museum in Marbella, which usually shows really bad art. They had the *Tragedy of War* series and the *Bull Fights* and I remember leaving this museum in a deluge of rain, thinking, how could you make art that profound, that brutal, that tragic, with that amount of pathos, while living in Marbella, in wonderful weather, surrounded by beautiful bodies and eating fantastic food? The contradiction between really that profound, emotionally messy art and those idyllic surroundings made me want to put these two things together in a painting. From that point, in about 1996, I started to paint incredibly beautiful blonde girls on park benches, trying to explore the motif of desire and the leisure space against a pictorial position which emphasizes some kind of horror, misplacement and anxiety.

Bregtje van der Haak (BvdH): What interests me about the paintings is that they have this rather ghostly utopian quality, which evokes for me a

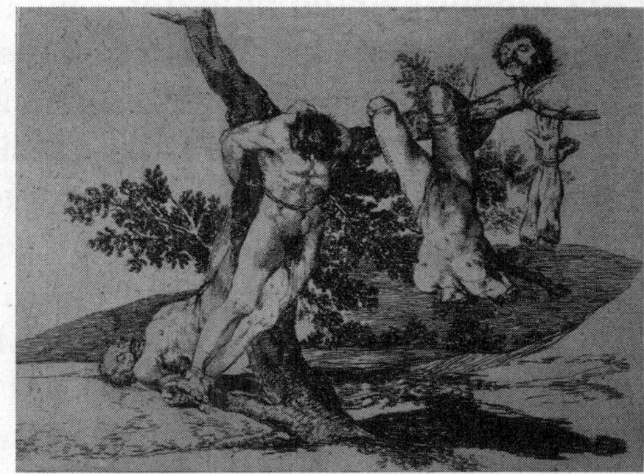

5.1   Francisco Goya, *'An heroic feat! With dead men!'*, 1810-1820

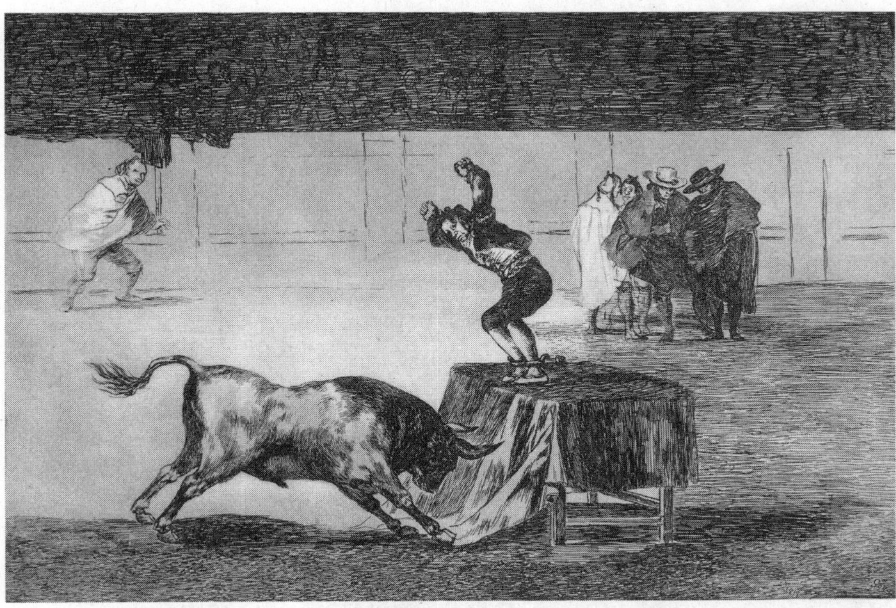

5.2   Francisco Goya, *Another madness of his in the same ring*, 1815

sense of loss, rather than a sense of happiness or success. I see a utopia that is no longer a collective idea of 'a better life', but a rather lonesome striving for personal fulfilment.

TS: The utopian idea comes out of Europe's theological crisis in the sixteenth century, and it's interesting that two of the people responsible for that were the humanists Erasmus and Thomas More, who fundamentally challenged the Church of Rome's idea of heavenliness with his book *Utopia*. I think theology is inescapable in image making, and religion relates in many ways to our contemporary idea of leisure or celebration or even holidays. I'm not making religious paintings, but I do try to make a kind of holy image. The scale of my paintings has this life-size aura and a kind of Catholic glamour about it. What I am interested in is the kind of universal notion that there is something better in life. I think we are now trivial economic beings, yet we still do believe in a kind of holy leisure lifestyle, just like people in the Middle Ages believed they might go to heaven. Unfortunately, nobody seems to ever get what they really want – it's just an image held in front of us to give us a reason to stick out the reality of being a bank employee or a car salesman, or during the Middle Ages, the banality of ploughing a field.

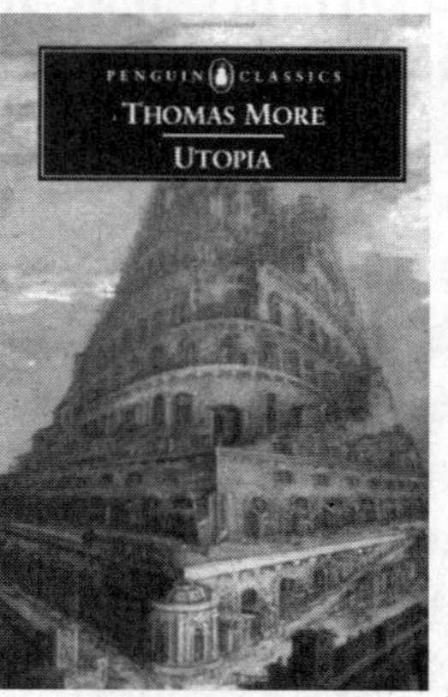

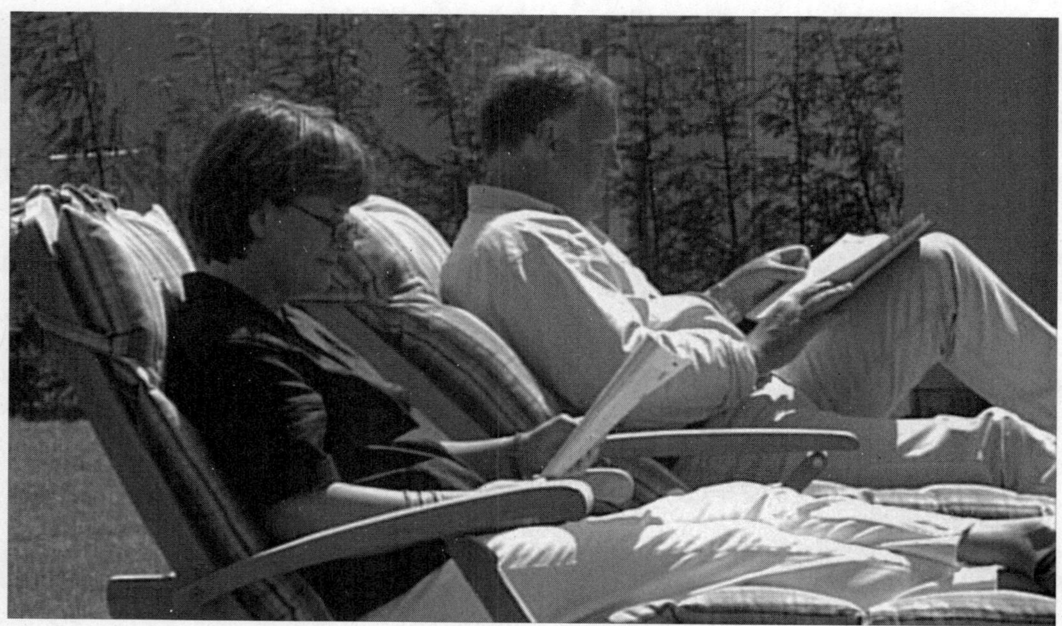

5.3   Bregtje van der Haak, *De vrijetijdszone*, film still, 2001

BvdH: I think it's interesting that you talk about leisure in relation to religion, because in rich, secular societies, the quest for meaning seems to be largely located not in work but in leisure space and time. It's at once very trivial and very profound. Like religion, leisure promises to fulfil, to put an end to longing. But it never does. In *Cocaine Nights* J.G. Ballard describes a perfect leisure community on the Costa del Sol that under its neat surface fosters all sorts of crime, drugs, kinky sex and transgressive behaviour, just to feel alive and not die of boredom. In Zeeland, tourists don't just go for beach walks or bungee jumps. They do rebirthing with a therapist in a swimming pool, forever looking for that ultimate experience, which they will never find.

RW: I recognize that. For me the paintings are about a universal desire to be free of work and worries, but also about the limitations and the problems with that. I know that once I am on that palm beach I will be bored and unhappy. I think your paintings convey that ambiguity very well, and I would like to talk about how your images function; you once told me you want them 'to resonate'.

TS: The idea of resonance is important to me, because there is something very primitive about a flat image. Cave painting was flat, and modernism

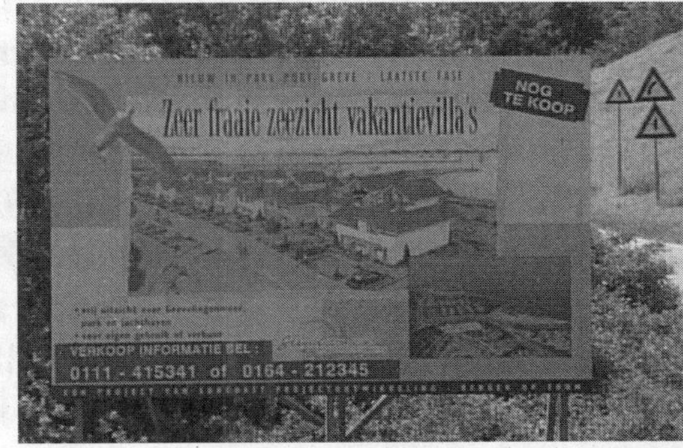

5.4 Bregtje van der Haak, *De vrijetijdszone*, film still, 2001

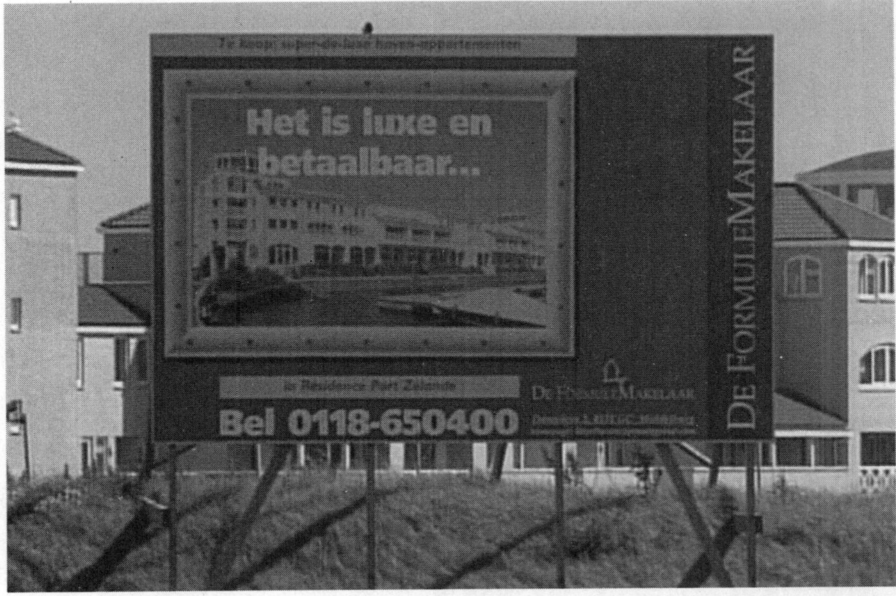

5.5 Bregtje van der Haak, *De vrijetijdszone*, film still, 2001

tried to return to the idea of some kind of primitive
spirituality in the surface. Without sounding too
much like an old Romantic, I love that notion, and
when I first tried to do it I felt it was almost embarrassing to talk about it. Resonance in a painting can
work in many different ways. One is the pictorial,
visual way, which has to do with a sinister overtone
in the light, but there is also another way, about
how an adult may hold a child for example: is it a
gesture of affection or a gesture of aggression? I
hope to leave these things open to interpretation,
so that what could be perceived as an idyllic notion
of a Sunday afternoon could also be interpreted as
a very problematic set of relationships by another
person. I'd like my paintings to be like sitting on
a sunny beach while watching a video of *Apocalypse Now*. At the end of that film you see these
fantastic stills of palm trees and they just explode.
For example, if I make a painting of two figures
running into the sea with a kind of atomic light, it
can be read at different levels: when you read it on
a shallow level, you just get a joyful image, but I
hope you also see the horror, the facelessness, the
lack of individuality of the figures. If I do my job
well, these polarities somehow find a contradictory
imbalance within the image that shifts it away from
the usual way you might locate it. That's where a
painting starts to work and in fact it only becomes
a painting when the facelessness is not just of the

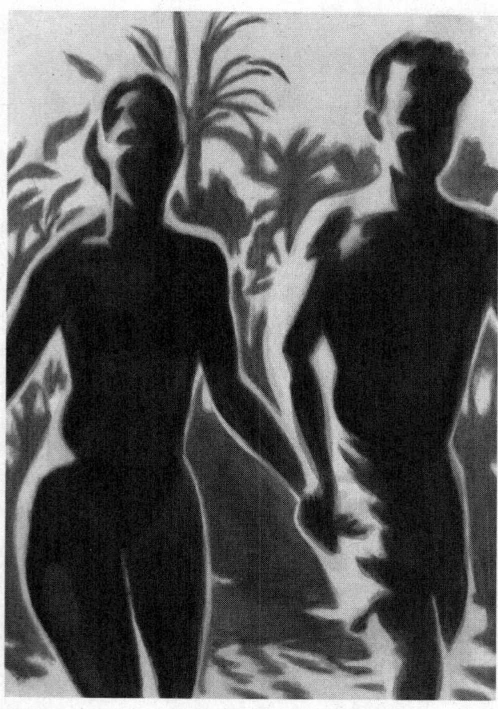

5.6   Tim Stoner, *Long Haul*, 2000

figures, but of the whole image. That's what I mean by resonance.

BvdH: I would tend to read the paintings in a more political way, namely as a very timely expression of the loss of collective goals and meaning. Living in Europe in an era of global capitalism and economic prosperity, all conflicts seem to have been resolved. Yet there is a sense of loss as well, not quantifiable, hard to describe, related to the unravelling of the social fabric. For all our connectedness through networks of fibre-optic cable, we are in many ways less connected than ever.

TS: My problem is I can't relate to any of those things without thinking in terms of images, which is probably why I am a painter. I don't wish to be moral in terms of dictating some kind of agenda, but I do hope to propose moral questions with my work. If we're talking about global capitalism, or the end of the Cold War, how do you reinvent ideas through painting? Thatcher said there was no such thing as society, only the individual. For me, making these images is important to reaffirm that we do have communal ideas, desires and sentiments.

BvdH: So that's all there is to it? Communal desires are now about being stranded on a beach with a coke and a Nokia phone?

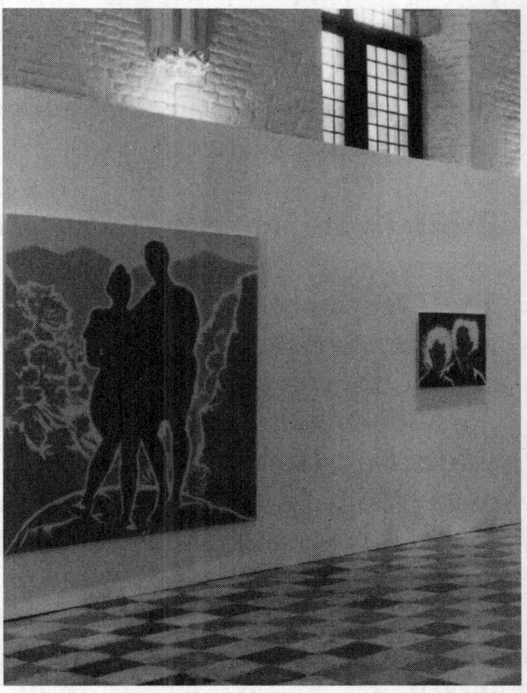

5.7   *The Lesiure Society*, Tim Stoner, installation view, 2001

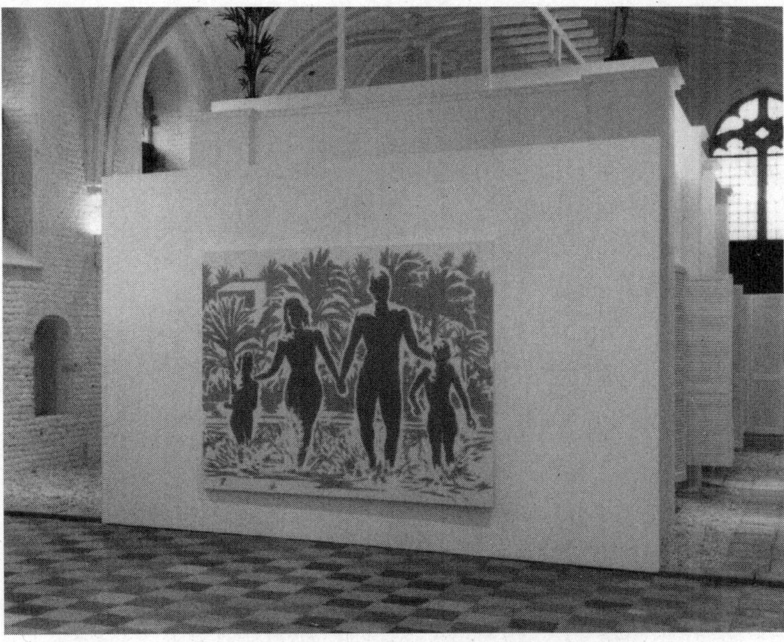

5.8   *The Lesiure Society*, Tim Stoner, installation view, 2001

TS: For me no, and that's why I make paintings about it. I am not just an observer; all this stuff applies to me as well. Ten years ago, I thought painting was all I ever wanted, and now I'm saying I also want to spend time in Marbella doing nothing. You can probably ask any man in the street and he might say he'd want the same thing, although I would love to be proved wrong about that. I think the simplicity of consumerism is terrifying in its pursuit of happiness through material desire.

RW: But I think you play with that in a very clever way. The paintings are political without being politically correct.

BvdH: Or are you saying they are ironic?

TS: If they were ironic, I could probably paint anything. I could turn up in my studio tomorrow and paint a camel or a still life, but I don't. I accept my sentimentality and my emotional response to the things I paint. Ironic painting, for me, is devoid of any kind of emotion. If there is any irony in my work, it's stamped on by pathos; it's chained down. When I paint a folk scene or some dancing girls, I really mean it. When a painting proposes a moral problem but does not give a moral answer, then that is something very volatile, but important. Hundreds of good paintings in history do that, but

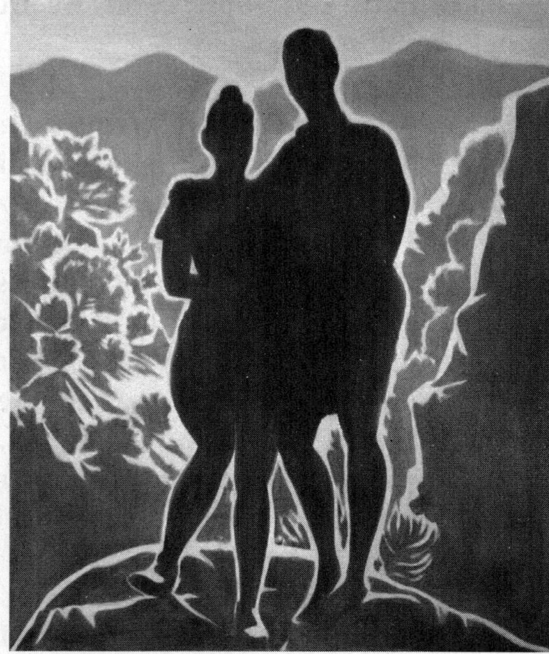

5.9 Tim Stoner, *Sierra*, 2000

the question is: how do you do it now? How do you make a painting that has that dual proposition? As an artist you're not just dealing with politics with a big P. You're dealing with very subtle nuances that might not be noticed at a glance, and that is where things get really interesting and critical. I know where my work plays with some kind of political agenda. I've had people read it as overtly heterosexual, or like some kind of right-wing lifestyle catalogue. But I don't see it like that. It's just there; it's the social history I live in and therefore the only option. In this image-overloaded society, the reason to make a painting is perhaps to slow it all down a bit, not just throwing another transparent media image back into the pool of that information circuit, but making something that has a different integrity, a different speed, a different pace and a different surface – and that for me is the reason to keep doing it. Every so often this conversation about painting being dead comes up, but I think that transculturalism and the way images move around now create a new situation, and all of a sudden anything seems possible as a painter.

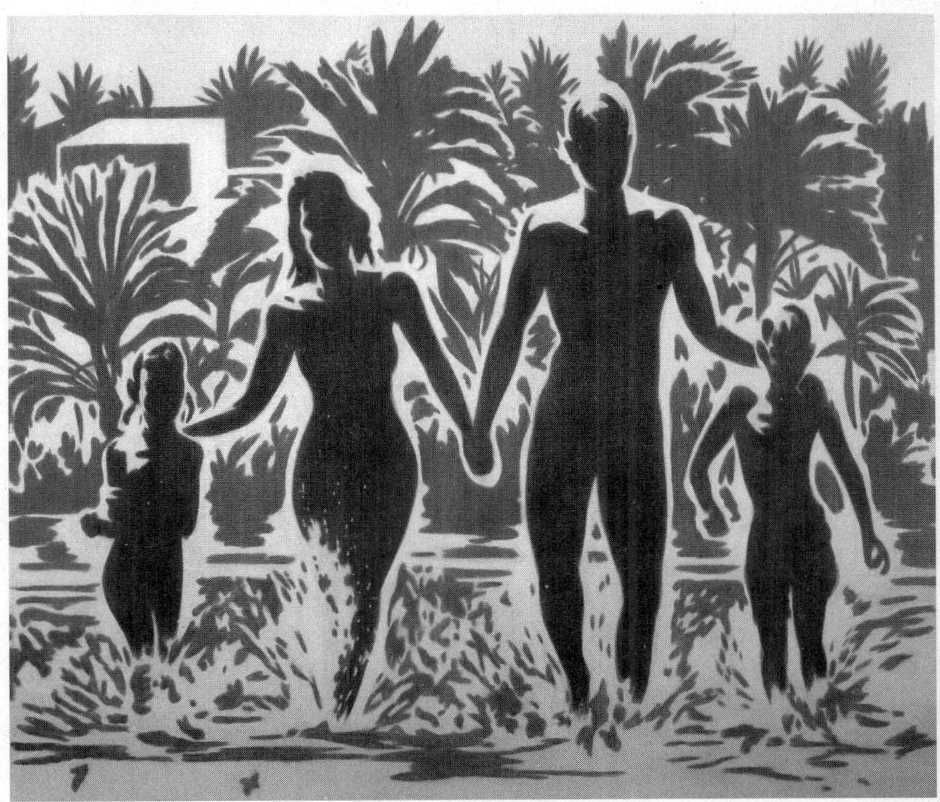

5.10 Tim Stoner, *Development*, 2000

Rutger Wolfson: Art in Crisis

1.

On the corner of a run-down street two women strut their stuff. Hot pants, lots of jewellery and *lots* of attitude. Two men in a cabriolet try to pick them up. They don't get very far. 'Get your black ass in my car, and let's get it on, bitch!' 'I ain't getting in your motherfuckin' car; you motherfuckin' broke-ass nigger!' Just as the men's advances threaten to turn into a brawl, an incredibly long limousine pulls up. A white man gets out and begins to dance, à la Gene Kelly in *Singing in the Rain*. He is joined by swaying dancers in too-small bikinis – and the mood changes from humorous to absurd. Then, as you realize the women's faces are slowly changing into the dancing man's grotesque grimace, it all becomes repugnant. What started out as a parody on rap music videos soon degenerates into a surreal nightmare. Sexy women's bodies – all with the same, terrifying man's head.

This clip, *Windowlicker*, was made by Chris Cunningham for music by Aphex Twin, otherwise known as Richard D. James: the man dancing like Gene Kelly. The first time I saw it, late at night on MTV, I had to reassure myself that nothing had changed. The clip is disturbing and unsettling. Not just because the images themselves are alarming, but also (especially) because they hold a mirror up to MTV and its viewers. Aphex Twin's electronic music is so abstract that it would not normally be played on MTV. However, the clip soon became a hit and was broadcast often. As a result, MTV's usual fare was further undermined by the venom of the *Windowlicker* parody.

The plastic arts have met their competition, for *Windowlicker* easily measures up to work by the very best video artists. The clip is no one-off – it is just one example of the broader development that is the focus of the publication *Kunst in Crisis*: the fading of the distinctions between artistic disciplines. The frontiers are pushed back from both sides. Music video makers are inspired by art, and vice versa: video artists take inspiration from music videos. A similar mutual influence is apparent on the boundaries between art and fashion, film, design and the new media.

This cross-fertilization is interesting because it not only blurs the distinctions between different artistic disciplines, but also those between 'high' and 'low' culture. The term high culture denotes disciplines considered to be of artistic importance, such as the plastic arts, architecture, literature, the theatre, dance and opera. Low culture refers to cultural expressions deemed to be of lesser artistic interest, such as fashion, advertising, music videos, new media and design. In addition, producers of low culture often work on commission, and its products are usually made and distributed in large quantities. An artist is autonomous; a painting is unique. A music video maker works on an assignment; copies are worth as much as the original.

The disappearing frontiers and the continuous cross-fertilization between artistic disciplines have called the distinction between high and low culture into question. Art and its related disciplines have grown towards each other to such an extent that it is sometimes difficult to differentiate between them.[1]

2.

This osmosis between 'high' and 'low' is clearly perceptible on the dividing line between the plastic arts and fashion. A number of fashion designers have a highly conceptual approach. Their designs comment on the traditions and conventions of their craft, much as many works of art comment on art itself. The conceptual perfume devised by designer duo Viktor & Rolf is just one example. A perfume is a label's status symbol: a label has to be very well-known in order to launch a perfume. Moreover, developing and introducing a perfume is phenomenally expensive. But, once launched, a perfume is highly lucrative. In other words, you have to

[1] One could also speak of a blurring of the line between two artistic disciplines with a more or less contemporaneous standing. One example from high culture is the blending of visual art and architecture. This is a primarily

invest money in order to make money. Frustrated with the harsh laws of commerce, young, relatively unknown and penniless designers Viktor & Rolf launched an imaginary perfume. The perfume was no more than an idea, packaged as an advertising campaign. The campaign was so realistic that you could have believed it to be genuine. The only clue that something was up was the model, who had collapsed after smelling the perfume. Viktor & Rolf are now successful designers; they are also respected in art circles. Not surprising: they speak a language art understands. No one has any idea what their perfume smells like, just as nobody knows how the faeces the artist Manzoni canned and sold as *Merda d'artista* (Artist's Shit) in the 1920s smelled.

Fashion designer Martin Margiela is another example. He, too, has a mindset comparable to that of the artist. His first fashion shows were set up almost as guerrilla raids. The audience was invited to a street corner in the middle of Paris, where the models flashed by, showing the clothes. A radical alternative to the rigged-up and controlled shows put on by the designer establishment. Margiela has also rebelled against the (commercially motivated) quest for innovation. Instead of presenting a new collection every season, he continued to work on one collection.

intellectual exchange, which is no threat to the standing of either art or architecture. On the contrary, they seem to serve each other to great advantage.

For some time, he kept presenting the same – subtly altered – collection. And even now the clothes made by Margiela do not bear the name of their maker: the label is blank.

Such a strategic, and successful, use of fashion's mortal sins – a blank label is now a recognizable logo – bears comparison to the way in which artists have, often successfully, attacked the mores of art. The example most often cited in this respect is Marcel Duchamp, who caused an outcry when he exhibited a found bicycle wheel on a stool in a museum in 1913. His display of a *readymade*, a found object to which he had added nothing, constituted a blistering commentary on the artistic conventions of the day. Margiela's and Viktor & Rolf's critical approach to the unwritten rules of their profession is akin to that of Duchamp and many other artists. This has resulted in something of an artistic upgrading of fashion. Thanks to them, fashion has become more of an art.

Conversely, art has become more fashionable. Some artists are fascinated by fashion's glamour and imagination. Italian artist Vanessa Beecroft's performances are a notable example. Beecroft has models with the same physical characteristics, all wearing the same clothes, pose as a group. Thanks to her impeccable staging and choice of clothing, her performances look like perfect, hushed fashion shows. Dutch artist Madeleine Berkhemer's early work is also closely related to fashion. Among other things, Berkhemer designed elastic bands you wear over your ordinary clothes, around your chest perhaps, or around your arm. These elastic bands refer to the internal structure of clothes: the clothes are, as it were, turned inside out. But they also refer to fashion's brand orientation. The elastic bands all have labels attached to them. This label is superimposed on every other label, as the elastic bands are worn over clothing. Art making statements about fashion, as in the case of Beecroft's and Berkhemer's work, nourishes fashion through a form of artistic reflection that automatically makes fashion more of an art.

3.

This kind of two-way traffic of ideas, strategies and respect takes place within every discipline related to art. Increasingly, graphic designers now develop concepts in much the same way artists do – for example by making the normally hidden, underlying structure of design visible and incorporating it in their work. Dotted lines indicate the type area, or the words 'page number' are actually printed. Some designers add conceptual depth to their work by adding hand-written comments, which are then printed. Add to this the social engagement displayed by some designers, and it would seem that they are consciously striving to elevate their work to the status of art. This is not surprising, for, conversely, conceptual art is becoming more and more dependant on design. The more conceptual the work, the more the artist makes use of graphic design in order to present his or her ideas.

That the blurring of distinctions between artistic disciplines has become an important issue in the art world, is also evidenced the double position adopted by many artists. A bit of ambiguity is interesting. Many artists do something on the side: as web designers, for instance, or as cooks. The occurs in other disciplines. I know graphic designers who are also DJs, and art historians who play in bands.

4.

The present-day blurring of distinctions in art is more far-reaching than the long tradition of artists' fascination with low culture, demonstrated by (for instance) Andy Warhol and Jeff Koons. This fascin-ation has always gone hand in hand with detached critical reflection. In this tradition, works of art comment on low culture without really engaging with it. Jeff Koons, for example, created a sensation with porcelain sculptures bordering on kitsch. The innovation, at the time, was that the sculptures were presented in museums, pre-eminently the setting where kitsch does *not* belong. In so doing, Koons did not, as he himself provocatively maintained, make a statement about the beauty of kitsch, but about the values of high culture. However much media coverage his statement generated, ultimately it applied only to art itself.

This tradition illustrates the autism of art. Ever since art first required comment and explanation, because looking alone no longer sufficed to understand, art has been its own subject – more and more so. Art as 'an eye turned inwards'[2]: a work of art always expresses some statement about art. And so innovation has become a subtle play on the discipline's own traditions and conventions. An innovative work of art invariably is a reaction to another work of art, which in its turn forms a reaction to yet another work of art … et cetera. It is interesting, given this introspective bent, that art is now venturing beyond its own frontiers, in the name of innovation.

Joep van Lieshout's work is an example of this. His lavatories, bathrooms and kitchens are indistinguishable from design. He is considered to be an innovator, because he explores the outer limits of art. But whereas Koons underlines

[2] Expression borrowed from a lecture by Cornel Bierens, delivered on 22 November 2000.

the boundary between art and low culture, Van Lieshout crosses it. Van Lieshout's designs can be seen both inside and outside museums. They have been installed in places where they are purely functional, and where they have to hold their own against (other) design. This is an essential difference. Had Koons tried to sell his sculptures in kitsch shops for art prices, no one would have been interested. But Van Lieshout, too, keeps his distance from low culture. His work is not put into serial production. Nor is it sold in Furniture Heaven – it is, after all, art.

Its artistic status affords Van Lieshout's work protection. Its success is measured by artistic standards rather than by the criteria applied to design. It is important to Van Lieshout that his designs 'work' within an artistic context, as they are only really interesting as reactions to work by artists that have preceded him. He does not have to fulfil the prerequisites of mass production, or achieve high sales figures. His work may look like design, but it does not have to meet the same requirements. This is a kind of double standard often applied by artists venturing into other disciplines. At best, artists uses their freedom to contribute to another field, for instance by revealing omissions or taboos within that disci-pline's thinking. At worst, they make fools of themselves by producing work inferior to that of the discipline's professionals.

5.

Chris Cunningham is a good example of a professional whose work makes most artists' efforts pale in comparison. He began his career creating special effects for films – working on *A.I.*, for instance, before Stanley Kubrick's death. He achieved international fame through the music videos he made for artists such as Madonna, Björk, Squarepusher and Aphex Twin. He produced his first work of art, *flex*, in 2000. *flex* was not commissioned by a record company; it was made on the invitation (and funding) of powerful London gallery owner Anthony d'Offay. While he undoubtedly recognized Cunningham's genius, Cunningham's reputation – precisely as a video clip maker – probably was an additional reason for d'Offay to make this investment.

In making *flex* Cunningham had complete freedom. He chose, once again, to work with musician Richard D. James. The result was a video that is not much different from Cunningham's music videos. It contains more explicit sex and violence – strictly prohibited by MTV. *flex* may not be shown on television, or at festivals. Nor has it been released on video or DVD, as *Windowlicker*

was. Those wishing to see *flex* have to go to a
museum, for *flex* is art.

Since *flex* hardly differs from the work
produced by Cunningham for MTV, the distinction
between high and low culture appears to be somewhat artificial here. However, in this case the
distinction primarily has to do with commercial
interests. Anthony d'Offay wants to make good on
his investment. As a gallery owner, he can only do
this by introducing Cunningham to museums as an
artist; in hopes of selling his work. And so *flex* is a
good illustration of the spasmodic reactions of the
art world to the blurring of the distinctions between
high and low culture. The art world has two strategies: either it annexes, or it ignores. Cunningham
is an example of annexation: he has, as it were,
been incorporated in the art world. Cunningham is
no longer a music video maker who is better than
most artists; he now is an artist.

6.
Similar, but less successful, attempts have been
made to annex relatively new forms of visual
culture. Because of the common ground between
VJ-ing and film and video art, a few museums have
taken the bold step of showing the work of VJs.
DJs and VJs were more or less transplanted from
clubs to museums. However, the context of a club is

hardly that of a museum. As a result, the status of the VJ performances was unclear. The public did not know what to do: should they party, or should they watch and listen attentively, not touching anything? This demonstrates the problems museums have in understanding anything outside the gospel of art (and art history). Museums prefer to wait for low culture to become art, perhaps through an artist making some statement about it. Artist Rineke Dijkstra, for instance, made a video work about a club in Britain. Why set up a club in a museum if you can observe it from a distance, through the eyes of an artist?

Much the same is apparent in efforts by museums to somehow make room for the internet. The art world was among the very first to ascribe a role of great importance to the internet. As anyone with a computer can make a site, the internet is rightly seen as the one veritable universally accessible mass medium. Much was expected of art on the internet, but to this day it has remained unclear what the traditional museum can bring to bear. The art history perspective offers no starting points. VJ-ing and the internet lie beyond the traditional field of expertise of museums – and so most museums choose to ignore them. Perhaps then it will all go away.

This conservatism is also borne out by the approach of museums to disciplines traditionally more closely linked to art, such as (fashion) design. Fashion is almost always highlighted in one of two ways. Either from the perspective of the artist, whose work provides some critical comment on fashion, or by presenting designer clothes, preferably in the industrial arts section. In the latter case, the presence of ephemeral fashion in the temple of high art is legitimized by focusing on craftsmanship – as also demonstrated by the silver salt and pepper shakers exhibited in the same section. Of course a fashion designer is permitted to make critical statements about art, just as long as he or she works these into a piece of clothing. His or her ability to express criticism in a different way is ignored: that is the province of art.

7.

This annexing and ignoring hides a deeply rooted cultural pessimism. Both strategies are aimed at maintaining as much of the distinction between high and low art as possible, for fear of losing the universal values of art. In this line of reasoning, art – and therefore museums – exist by the grace of the distinction between high and low culture; the blurring of that distinction would plunge art into a crisis. For, it is argued, if the boundaries between high and low culture were to disappear altogether, art would go under in the torrent of images bombarding us day after day.

Paradoxically, however, it is precisely this protectionism that poses the real threat to art. To play the blurring of distinctions down as 'art testing its limits' is to relate the consequences of those fading boundaries only to art itself. The desire to keep low culture at arm's length merely deepens art's autism. Even when art *does* look toward other disciplines, in the end it only sees itself. The inability to approach the blurring of the distinctions between artistic disciplines from any other perspective than one of art history only drives art further back into the obscurity of its self-chosen isolation.

The desire to reduce the fading frontiers to a trend that will blow over stems from the fear that art will lose its autonomy. This autonomy is what sets art apart from low culture; it allows art complete freedom in developing insights into ourselves and the world around us. In this respect, art is often likened to science. Both would lose their legitimacy if they ever gave up their independence. Yet art and science differ in the ways in which they safeguard their autonomy. Science has rules that must be adhered to for something to be 'scientific'. In art, autonomy is coupled not to rules but directly to artists: only someone working in complete artistic freedom can produce art. As long as the artist engages with his or her own artistic tradition, any topic, any point of reference, any form is permissible.

Argued in such a way, museums exist solely to serve art and vice versa. However, the examples I cited above illustrate the need to modify the importance of the artist's autonomy. On, for example, the boundary between art and fashion, artistic qualities often have little to do with autonomy. Of course the level of autonomy is determined by the context in which someone works: the museum or the fashion label. But Viktor & Rolf's perfume or Martin Margiela's shows are no less interesting than the work produced by Vanessa Beecroft or Madeleine Berkhemer. The fact that the maker's autonomy is now an outdated criterion to judge art by is also born out by Chris Cunningham's *Windowlicker* and *flex*. Thus, museums cannot persist in claiming to be a setting exclusively for traditional artists. The more they continue to do so, the more untenable it all becomes.

But how can museums welcome low culture without at the same time proving themselves to be superfluous? After all, it is argued, you hardly need museums to see music videos, VJs or fashion – they're already out there: on MTV, in clubs, on the street. Fading boundaries appear to place museums before a dilemma with only self-destructive options: either pretend there's nothing there and so slowly fade away into obscurity, or embrace change and in doing so admit that the institution of the museum has become redundant.

The only way out of this dilemma is to let go, inasmuch as is possible, of the art history perspective. The autonomy of art should not be coupled to the artist, but to the museum. The museum should not primarily be a space for art and artists, but a space for ideas. In order to force art out of its autism, to let it truly come out of itself, museums have to examine what art *can* be. Professionals working in other disciplines need to be actively involved. In the process, museums can contribute to an investigation into the possibilities opened up by the blurring of distinctions between artistic disciplines, instead of secluding themselves on the basis of criteria founded on the very existence of clear distinctions. For to be autonomous and to have contemporary relevance, art has to liberate itself time and time again – even from itself.

<u>Valentijn Byvanck & Rutger Wolfson:</u>
<u>Museums and Contemporary Language.</u>
<u>Interview with Geert Mul</u>

1.
It all really started in art school, in Arnhem. After
I'd first painted and then worked with video, I
started working with computers in my final year.
There weren't any computers at the academy,
and so I worked with my own equipment. I didn't
exactly get a lot of response. Some people thought
it was pretty good, but I'd also be told that I might
be better off working at Philips. I graduated from
art school in computer animations and, what fascinated me more, programmed works. I wrote those
works in Basic, a very simple computer language.
One of them was called *life, death and bingo*,
which was a kind of slot machine with the three
little screens displaying life, death and bingo.
If you got three of the same, you'd have bingo.
Another work I made at the time was called *then
and now*, a monitor across which the word 'now'
ran until, at a certain point, it changed into 'then'.
Finally, the entire screen would be taken up by
'then', whereupon the row would shift and then,
like the timepiece on a digital clock, change into
a new image. Since this computer language only
allowed you to work with very simple fields and
I was hungry for images, I went on to work with
video and put the programming aside for a while.

6.1   Geert Mul, *then and now*, 1990

The question of whom or what I made the work for hardly concerned me. Neither did I think too much about creating work for any specific place. All I wanted – badly – was to make and show the work. I just didn't know where that should happen. The work didn't fill any niche, nor was there ever much interest. After art school I travelled: to the United States, Mexico and Japan. I lived in Tokyo for almost a year, and there I made a lot of audio recordings. Later, in the Netherlands, I used those recordings for my first videos.

Back in the Netherlands, I worked closely with Leon Dekker, a real painter I knew from the academy. We made virtual exhibitions. We built a space in a 3D programme on the computer and there we'd have joint exhibitions. I'd hang video stills on the wall and among those he'd put up his paintings. In themselves, they were quite good – or at least they were for a collaboration. It's old-hat now, but at the time it was still pretty fresh. In all, we made something like 15 exhibitions. We went around the whole alternative scene: the Archipel, all kinds of galleries, and artists' initiatives. We sold the prints, too. The whole idea was that it was a product that should do the rounds. We made multiples, which cost 100 or 150 guilders.

After 18 months I came to the conclusion that it wasn't really going anywhere. I could have spent

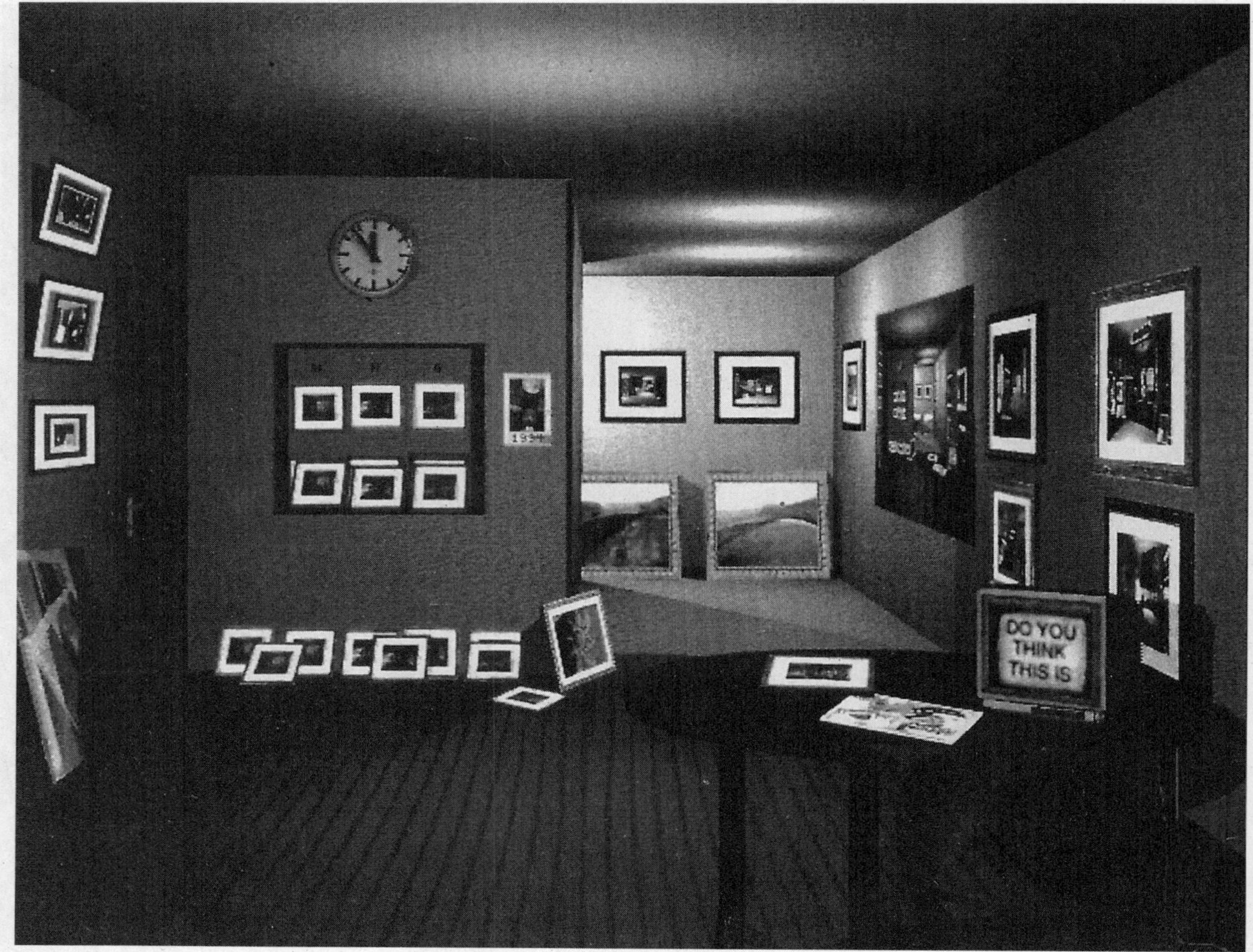

6.2 Geert Mul & Leon Dekker, *Virtual Expo*, 1994

the rest of my life making those things, but there still wasn't any real response. I could get rid of my work, it was selling, and I was allowed to put it up – although that usually cost money – but after a while I felt as if I were giving away sweets. You could spend your life doing that if you want to. It wouldn't upset anyone. But, every single time, it'd be your sweets and you wouldn't get anything back. Things may have changed by now, but at the time the alternative scene was still very closed. There was a network, but it was definitely a closed one.

2.

After reaching a dead end with the prints I took a part-time job with a video studio in Hilversum. That job gave me access to video equipment. I'd do some odd jobs during the day, and then I could work in the studio at night. Because of that studio I got some idea of the possibilities of video and also learnt to mix. It was in the beginning of house music, in 1994. I came up with the idea of mixing videos live. At the Rotterdam club Nighttown they were organizing 'nights' called *the Future*. I rang them up and asked if they'd be interested in a visual interpretation for the club. I told them I could do a light show with television screens: light and pictures at the same time. Well, there wasn't any other way to express it. I started doing that in collaboration with Titus van Eck. The work was great fun, and it made me financially independent.

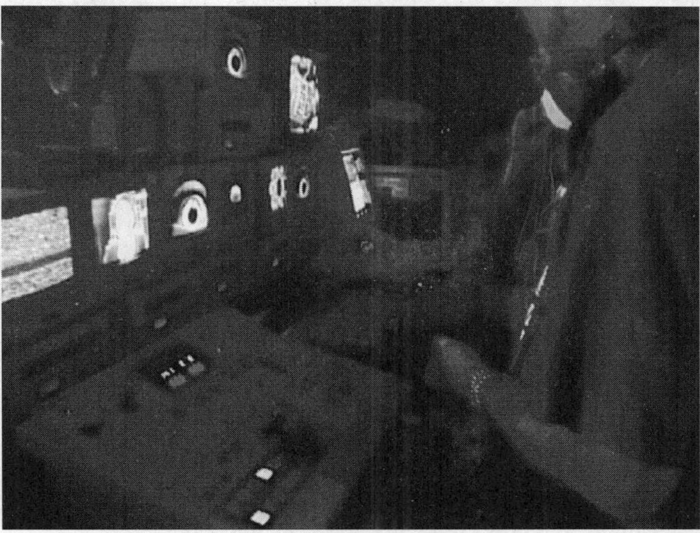

6.3   Geert Mul, *Lowlands VJ Set*, 1999

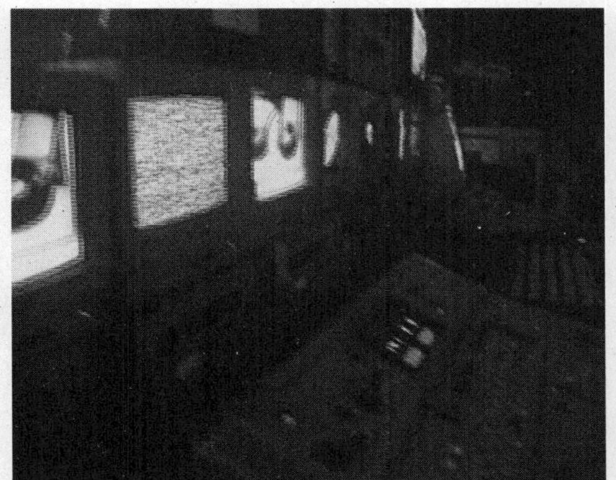

6.4   Geert Mul, *Lowlands VJ Set*, 1999

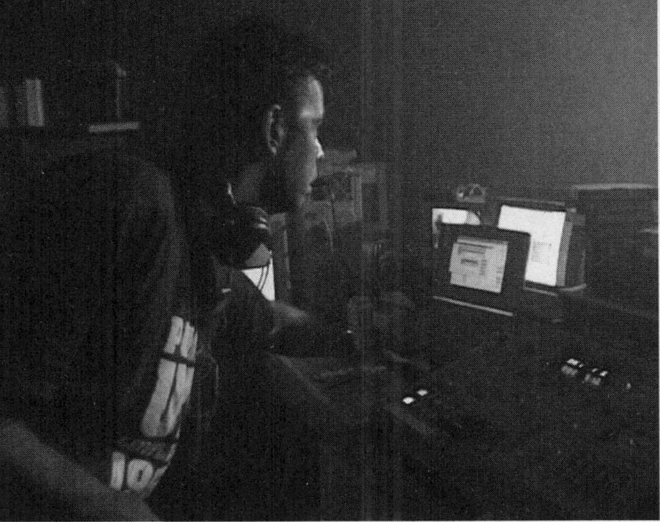

6.5   Geert Mul, *Lowlands VJ Set*, 1999

It also was very exciting to get close to music again. There was such a lot of new stuff going on: House, Techno, Jungle, Trip-Hop, Gabber. Perhaps the most important thing for me personally was that I could drop the idea that I had to exhibitions in a certain scene and build a career that way. I stopped looking for places to exhibit. I didn't have to bend over backwards anymore, trying to write up what I was doing within a certain jargon in order to get money.

One year later the work that was by then called VJ-ing had become hip. In the space of one year we made three television appearances, and as a result we were being asked for festivals. After a couple of years the visual arts, too, started showing an interest. Micha Klein became popular. It all started to take off. After a year or so I started running into people involved in the same things, in the same club, or the same magazine. We didn't really have predecessors. Sure, there'd sometimes be a projector in a club. But the idea of doing something with live mixing and music just suddenly popped up in my head – and I think that this happened more or less at the same time in Amsterdam. It had to do with developments in technology: at some point a video mixer becomes smaller and more affordable, and then it makes sense to throw it in the back of the car and take it somewhere. Of course, everybody does something

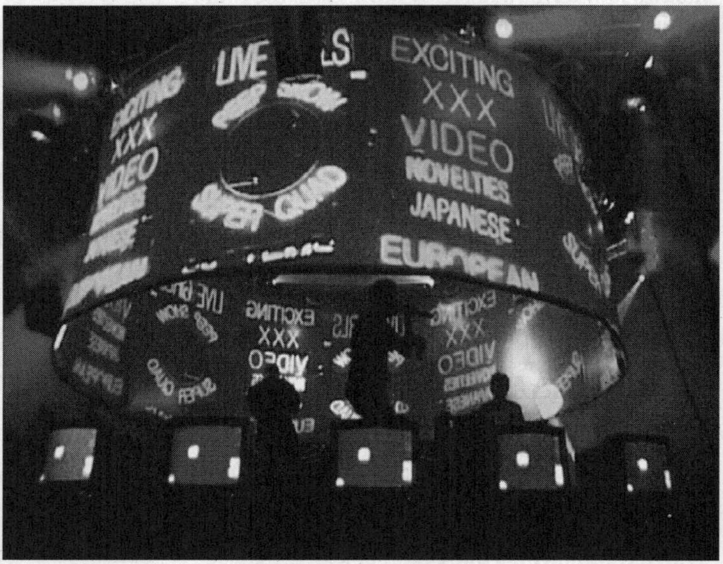

6.6   Geert Mul, *Lowlands VJ Set*, 1999

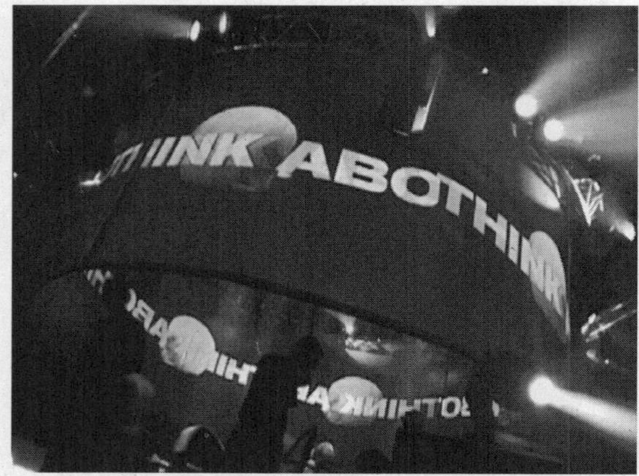

6.7   Geert Mul, *Lowlands VJ Set*, 1999

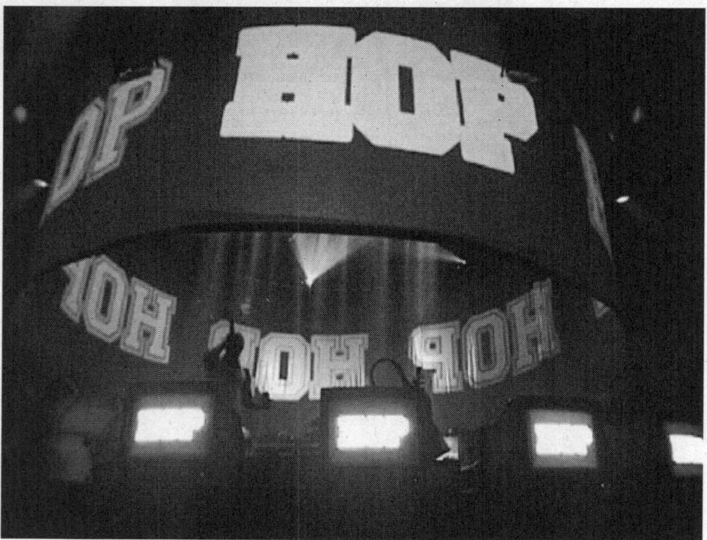

6.8   Geert Mul, *Lowlands VJ Set*, 1999

different. Ebo-man opted for making actual sets and is building a career as a performer and musician. Micha Klein went for the museum. Gerald Van der Kaap is a different story. As a designer and an artist he's been using video and computers from the outset. I think he's one of the most inspiring artists, because he handles his medium in a truly innovative way. In my own work, the collaboration with Gerard Koot has been very important. Koot combines a lot of know-how with a feel for technology.

I don't know whether the conjunction of music, space and people in those clubs in itself produces all that fantastic an experience. I also had a problem with the launch of the term VJ. In a club or at a pop festival it's all about the music, so in principle, as a VJ, you play second fiddle. It's not all that high-flown; you never drop off your chair in amazement at those gorgeous images. You're not supposed to, because you're there for the music – and the lights, decor and images all serve the music. I'd basically stand next to the lighting technician, who served more or less the same purpose. We work on a grander scale now, but in the beginning we just had three VHS players, a video projector and a mixer with which you could make all kinds of effects and stills and could synchronize to the beat. We weren't too bothered, especially not at first, about which tape

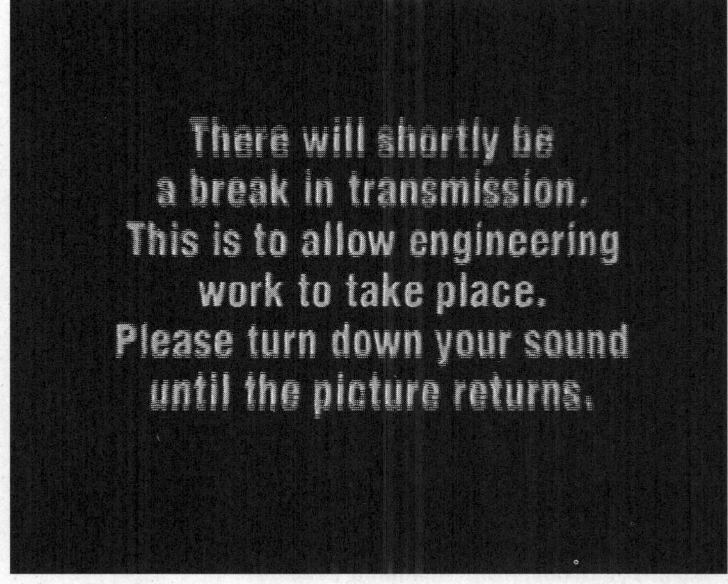

6.9   Geert Mul, *Lowlands VJ Set*, 1999

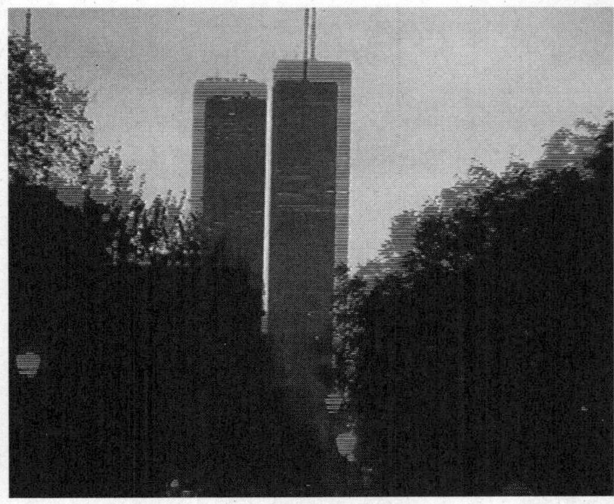

6.10 Geert Mul, *Lowlands VJ Set*, 1999

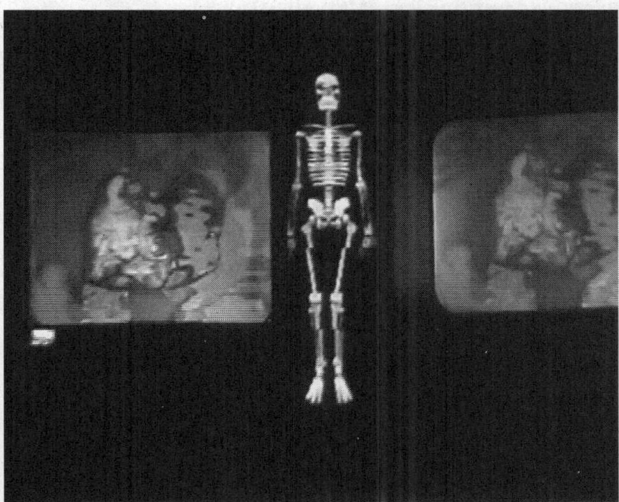

6.11 Geert Mul, *Lowlands VJ Set*, 1999

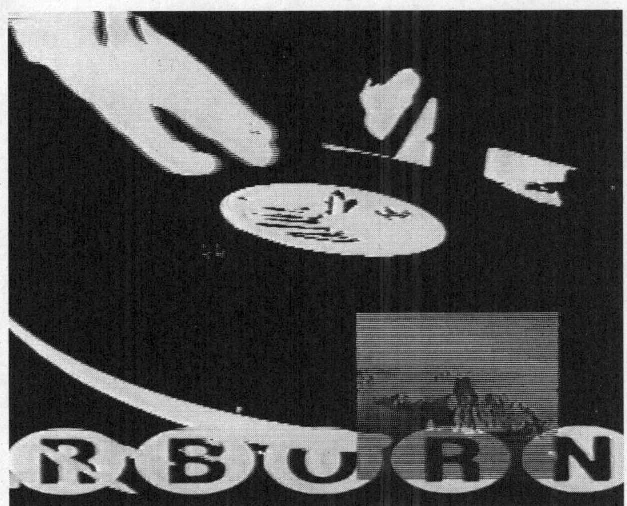

6.12 Geert Mul, *Lowlands VJ Set*, 1999

was in the player, because there'd be so many layers with a twist added on that we'd end up with very abstract images, and as far as we were concerned it was all about rhythm and colour. You get fed up with that after a while. The VJ always has to wait and see what the DJ will play and then select the images to go along with that.

Working in clubs has influenced my thinking on space. The space is different every single time, and so are the people and the music. You wonder what on earth those people are doing there. What have they come for? You try to figure that out and make your imagery respond. In our collaboration Koot and I have come up with a different visual solution for each space – one that was specifically grafted onto that space. In the video scene, everything is standardized. In principle, everything is 3 by 4 metres and hangs flat on the wall. If you want to diverge from that, you have to have a pretty good idea of what you're doing. Now that I've made my first interactive works, I realise that I've learnt a great deal from responding to spaces and people in clubs. Therein lies the core of the work, and that truly is something I hadn't realized before.

3.
If you were to ask me where the most interesting developments are taking place now as regards

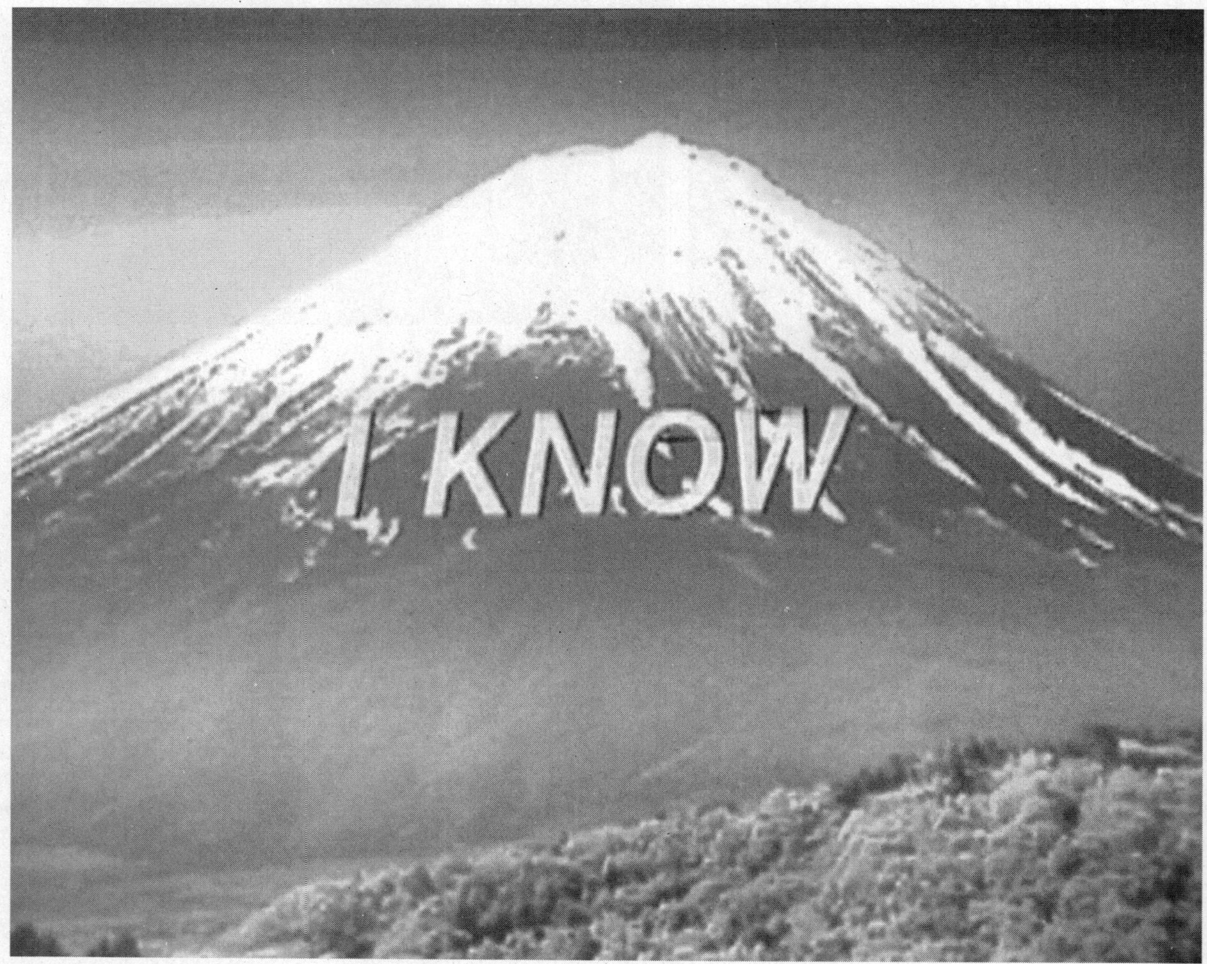

6.13 Geert Mul, *Generating Live*, video still, 2000

content, I'd say in the game industry. That's where the newest, the latest developments are in the visual, the interactive, language. You wonder why the visual arts, at a certain point, move to a more abstract level? To put it bombastically, I think there's a major change taking place. Evidently, all kinds of processes are crying out for a new language, new descriptions of our reality. That language shouldn't just be dynamic, but also be able to generate intelligent change. Before such a language can even begin to develop, you have to be capable of deciding what you want and be able to organize it. The cultural worth of technological changes underlying such processes is greatly underestimated, especially in the visual arts. That reveals itself in Dutch art collections, which are limited to a very specific Western European vocabulary that rejects all kinds of things: technology, sexuality, symmetry, monumentality. The chosen path is that of an absolutely individual expression and a highly abstract notion of purity – at the expense of contemporary complexity. I think it's important that a museum of modern art reflects the current language, and that's something modern art museums definitely do not do. They prefer to reflect upon a high point of painting from the Golden Age, whereas at that time they *did* paint and work in the language of the people. The greengrocer's sign was not too far removed from the painting by the artist, who used the

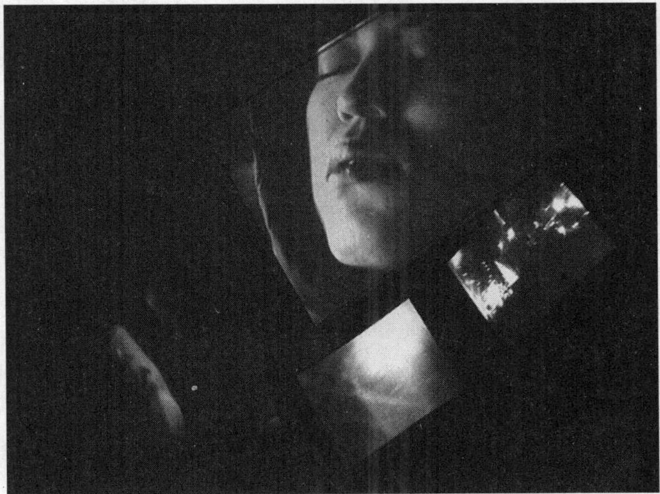

6.14 Geert Mul, *Generating Live*, installation view, 2000

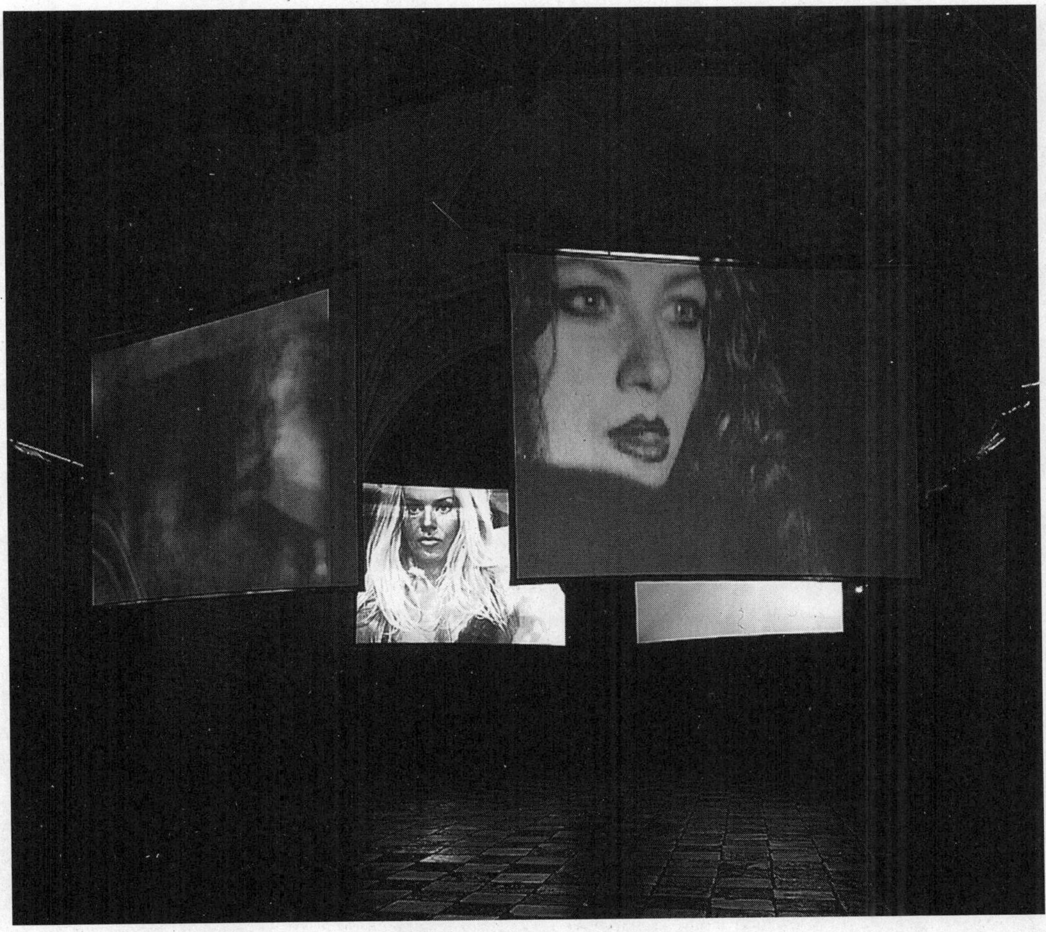

6.15 Geert Mul, *Generating Live*, installation view, 2000

same vocabulary and the same medium, but in a different way. There's no reason a museum of modern art shouldn't have 75 percent of its focus on the new media: film, photography, video, computers – whether or not combined with rediscovered older media like music, design, fashion and so on. To me, such a situation should be self-evident, as it would be a reflection of contemporary cultural developments.

It's also a problem of class. The museum represents the perspective of the moneyed class, which has always been crucial to the development of culture and art. The influence of this moneyed class on the cultural domain is tapering off. The museum can continue to present that perspective, but the relevance of this perspective itself is waning, and that is visible and palpable. At the same time, it is insanely difficult to make the transition to another language, another vocabulary, another target audience: the people will be coming from an entirely different circuit. You can see that on one of those 'museum nights' at the Boijmans Van Beuningen Museum. They have to organize something hip, somebody has a cousin who can do something, then it's too expensive and they have to make cutbacks, and as a result there's something hip fluttering about, and everyone concludes that those video things just haven't matured yet.

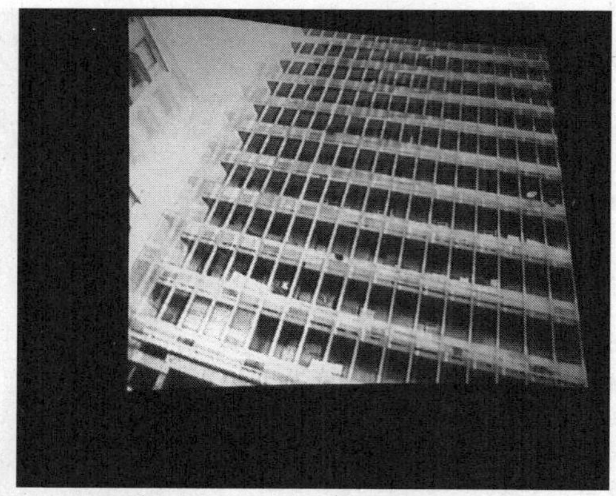

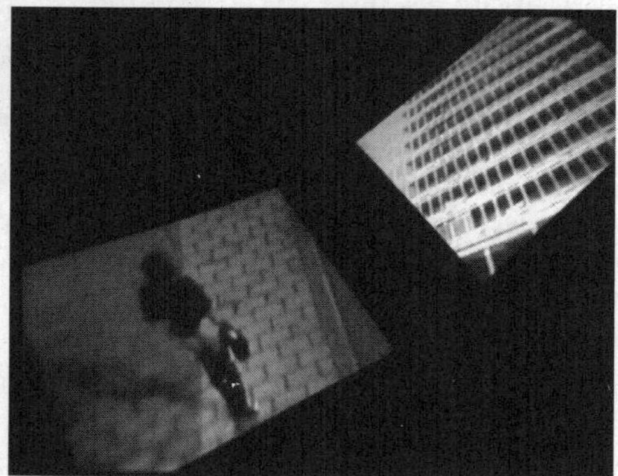

6.16 Geert Mul, *Generating Live*, installation views, 2000

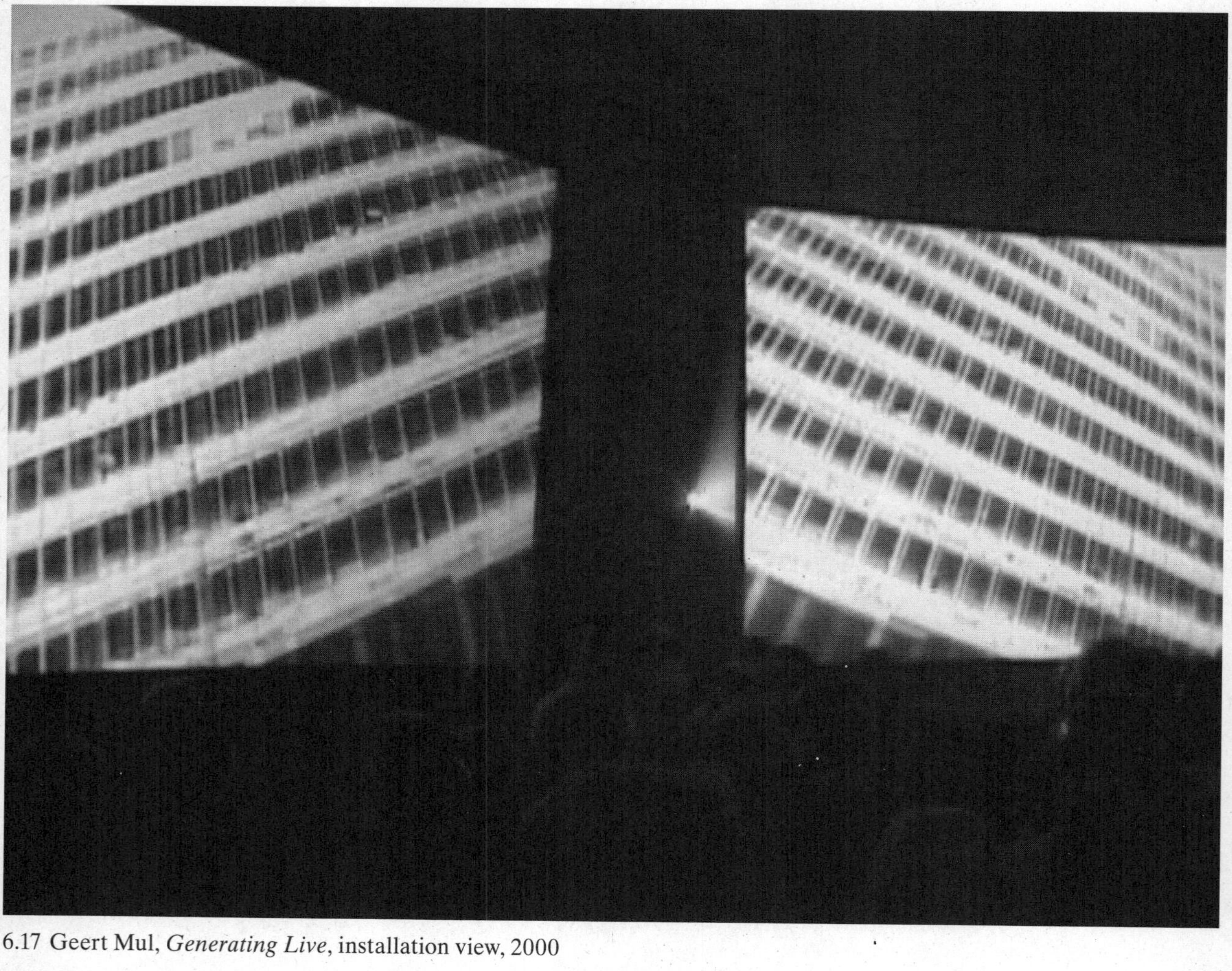

6.17 Geert Mul, *Generating Live*, installation view, 2000

Of course I could present my work elsewhere, but museums are important because of their funds and their power. That's why I challenge them. The museums have the room, the money and the support for projects. Recently, during the Rotterdam Film Festival, I saw some works at the Boijmans Museum that I thought were very nicely produced. Good space, good sound, very minimal means. It doesn't cost shit. When you talk about music venues in the Netherlands, you're referring to Paradiso, Nighttown and a couple of other places. When you talk about the visual arts, then there's Boijmans, the Stedelijk and probably a few more. That's where the facilities are; that's where state funding goes; that's where it's possible to show things. So why can't we go there on Friday nights to see films, watch videos and attend concerts and performances?

The reason I have always had and shall always retain a soft spot for the official art circuit is the level of the discussion conducted there. The dialogue, the discussion, is after all what it is all about. That is an important distinction to pop, which is always about the icon and the individual.

4.

Gradually, my own work and the work I do on commission have grown closer to each other.

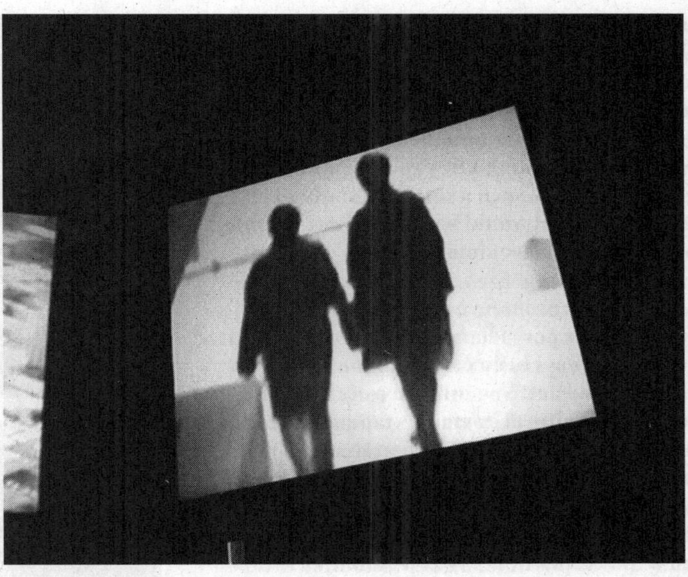

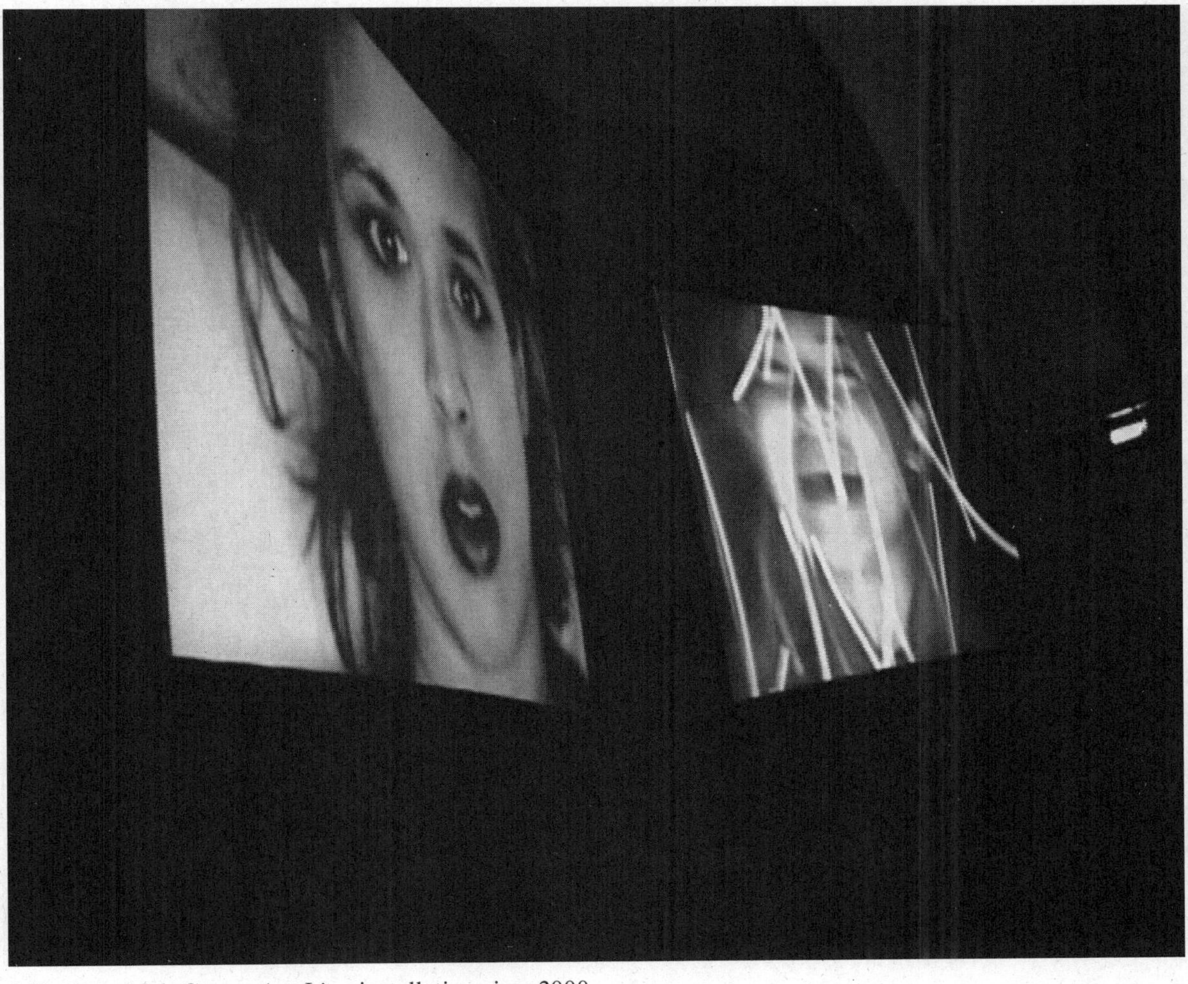

6.18 Geert Mul, *Generating Live*, installation view, 2000

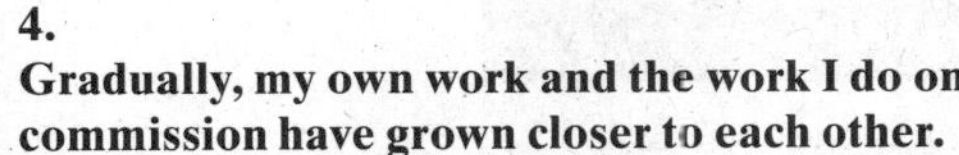

6.19 Geert Mul, *Generating Live*, installation view, 2000

Partly because I got back into music and through the things I discovered while mixing live in those clubs. The sets lasted something like six hours, so it's in no way about absolute beginnings and endings in images anymore, and your concept of time changes as well. They're *streams* of images, really long sessions in which you discover stuff through imagery, develop a feeling for rhythm, for colour. At a certain point you'd start getting an interaction. I'd use things I'd discovered in the studio in the club, and apply stuff from the club to my videos. The programmes I made after art school were really sort of dynamic works with fixed, repetitive patterns you'd switch on and which would then start running. At the time, I abandoned it because I wanted to do more with images and couldn't achieve, technically, what I was after. Eight years later, when the programming and the work with video converged again, it *was* possible. That was thanks to a new generation of computers on which I developed a method to programme or script video images rather than edit them.

Through working with music and computers in clubs I became ever more involved with the formal aspects of sound and image. This thinking about the role and meaning of technology is what sets me apart from the visual arts scene. The intrinsic position of technology as a given

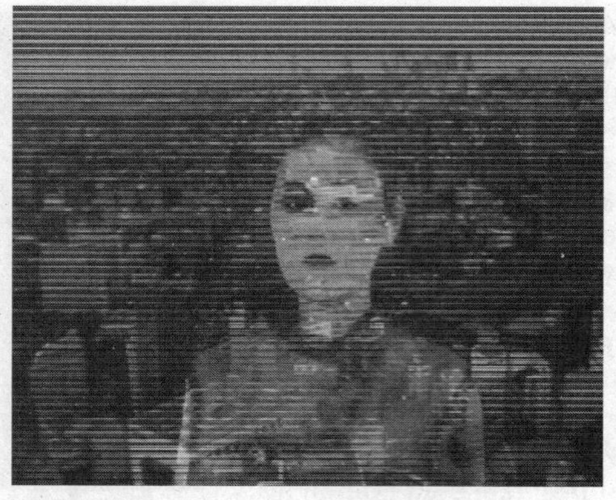

6.20 Geert Mul, *Tokyo FX*, video still, 1995

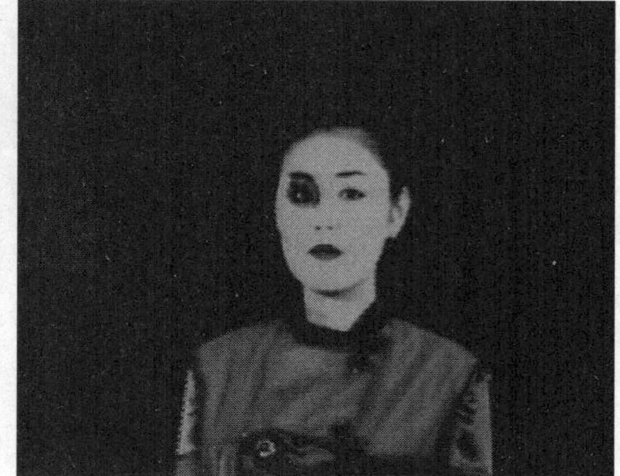

6.21 Geert Mul, *Tokyo FX*, video still, 1995

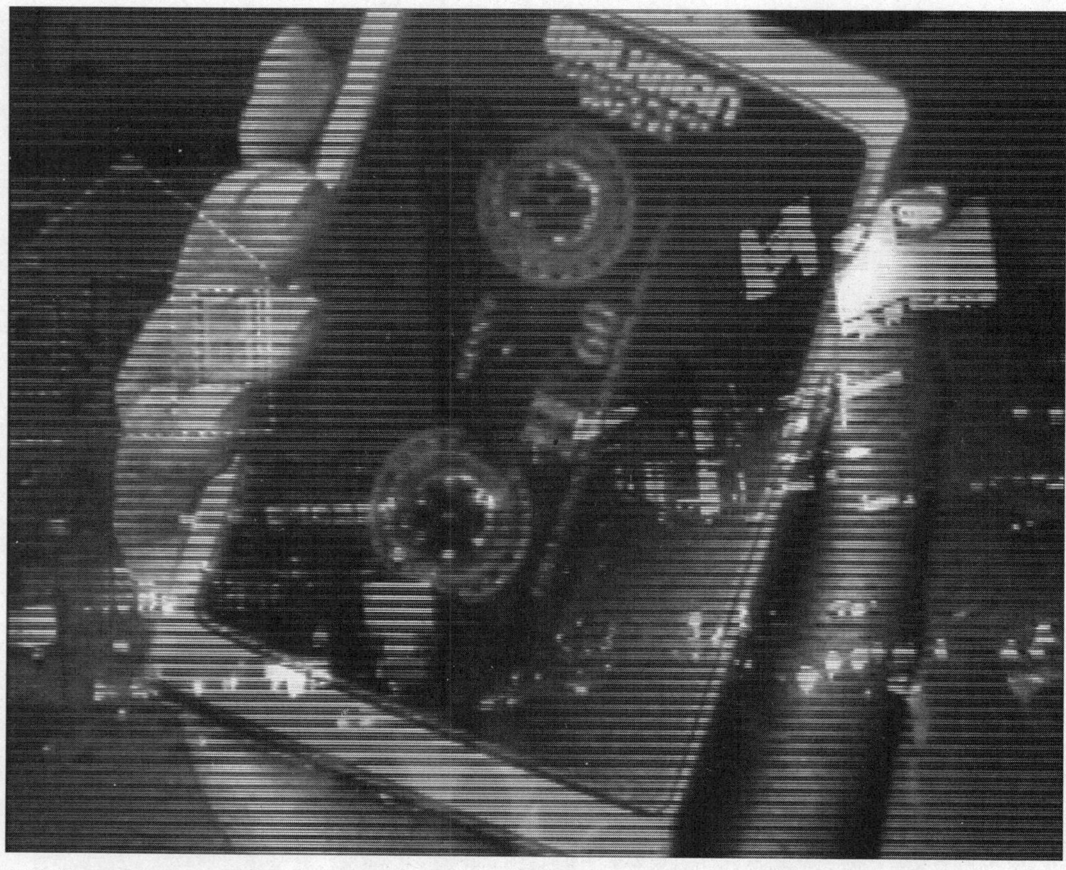

6.22 Geert Mul, *Tokyo FX*, video still, 1995

has been banished from art history. I'm often criticised in art history circles for not making any decisions; I think that's because, in my work, you make different kinds of decisions.

The undervaluation of differing decisions, new methods and possibilities is part of the history of technology. First a computer is built, and then it takes people 20 years to become aware of the consequences. When I first started art school the computer was still relatively new. The first experiments with interactive systems were very abstract, but a little later they were being incorporated into games like 'text adventures' that didn't follow a set pattern, but would take unexpected turns as a result of the information users fed them. Then, later on, you had the first experiments with interactive film. But that wasn't going anywhere. The viewer was dished up a film and then could select the plot, or at a certain point the image would be frozen and the viewer would be allowed to choose between a, b or c. These novelties were presented at technological events or audio-visual festivals. Everyone sensed it didn't amount to much. People didn't like the interactivity and so, for the next five or six years, the idea was dropped.

After the first, experimental, stage in the 1960s and '70s, all attention was directed toward marketing the computer. Instead of further

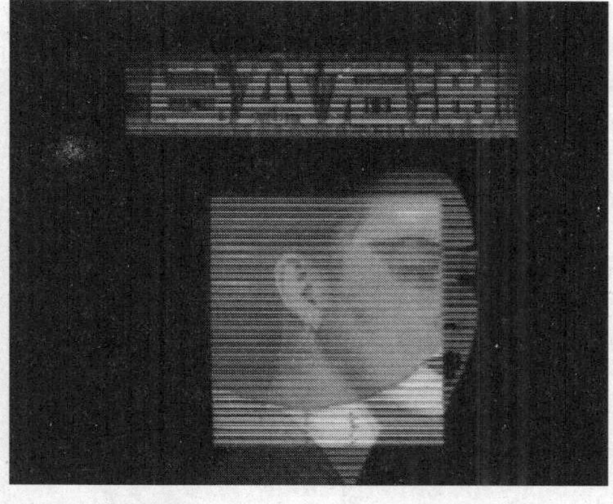

6.23 Geert Mul, *Tokyo FX*, video still, 1995

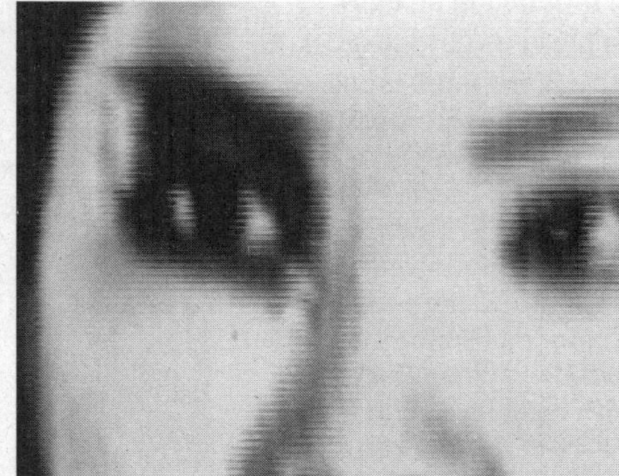

6.24 Geert Mul, *Tokyo FX*, video still, 1995

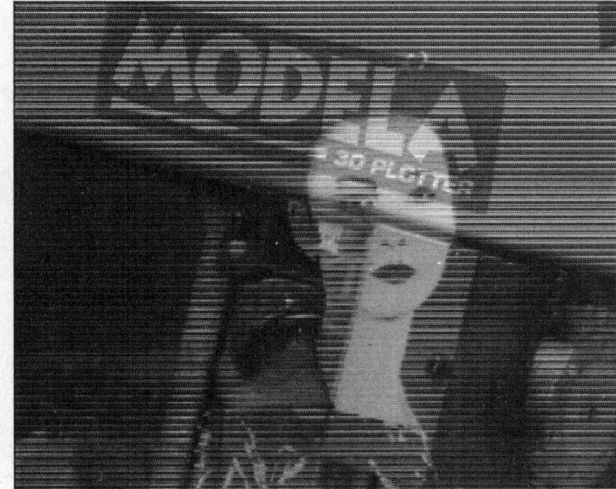

6.25 Geert Mul, *Tokyo FX*, video still, 1995

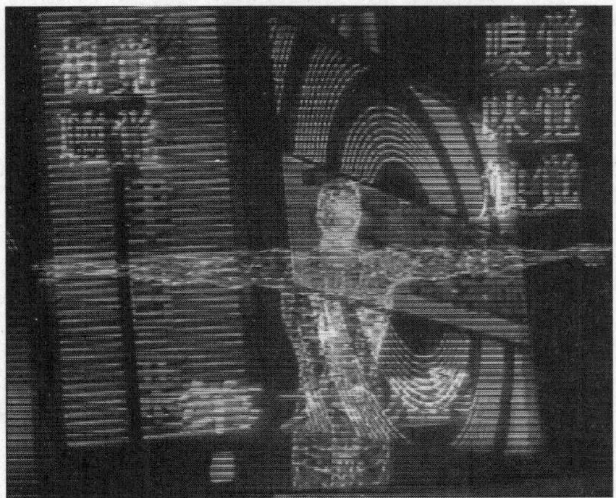

6.26 Geert Mul, *Tokyo FX*, video still, 1995

exploring its possibilities, the computer was attuned to the experience of the people who'd have to work with it. The computer was programmed in such a way that it could, for example, simulate an analogous editing system. It's called NLE, Non Linear Editing – only it just as linear as a tape, since the system uses a time line. You have to really over-tax the computer in order to do that, because if you let it go, it starts to generate, to mix, to make combinations and structures – that's what it's good at.

We've now come back to the possibilities that were already recognized 20 years ago and have since been propagated by the underground computer nerd scene. Their influence is only just starting to manifest itself on a societal level, in the interest in interaction and the letting go of the linear narrative. Actually, in the public sphere you can see how this type of interactive patterns is constantly being used to spread information on billboards, in road markings, traffic signs and advertising. The public realm is a dynamic given. If you can let your flow of information communicate with the public sphere and the people in it, you have a 100 percent gain. At Eindhoven Station they now have an information system that, if a train is delayed, can run a commercial for a florist within walking distance – where you can quickly buy a bunch of flowers. That is a new form of spreading information.

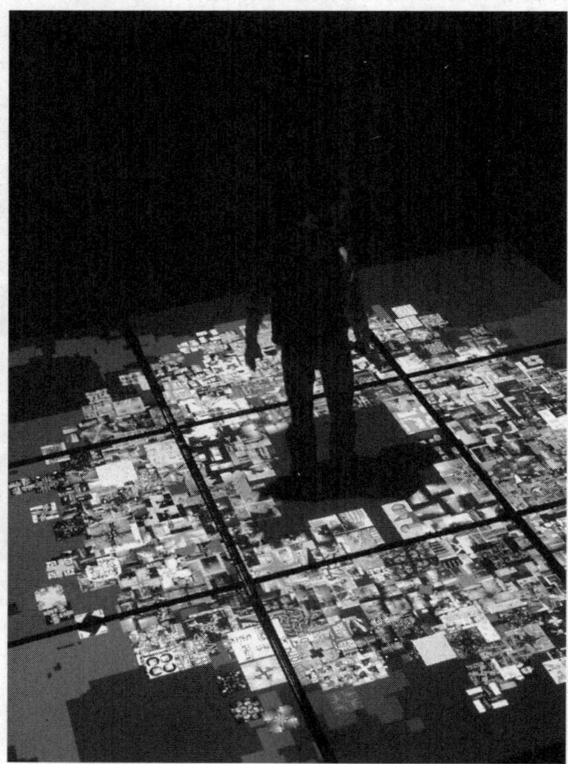

6.27 Geert Mul, *Babel*, installation view, 1995

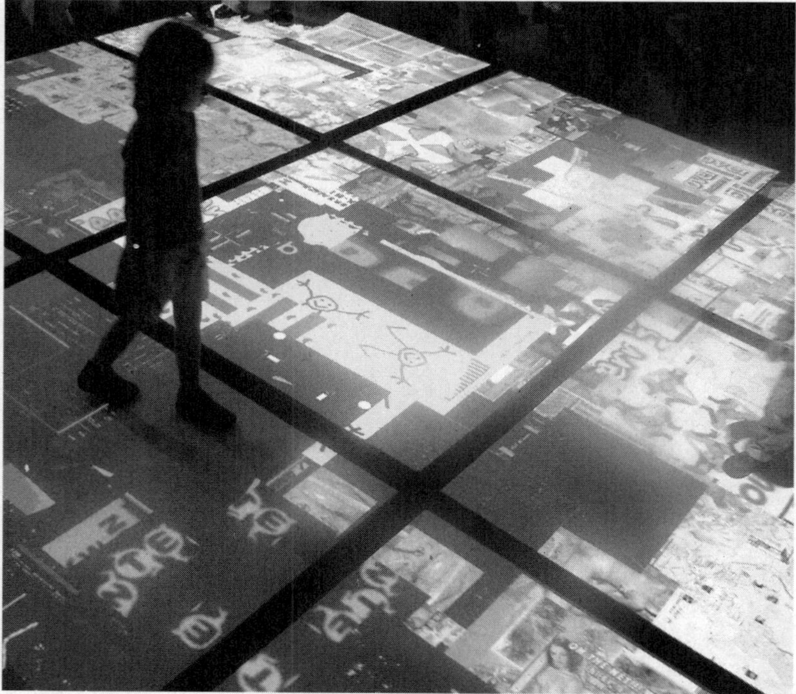

6.28 Geert Mul, *Babel*, installation view, 1995

The scepticism about interaction often stems from delineations and definitions that no longer apply. This is about a language we hardly speak, and that makes it incredibly difficult to communicate about it. Of course, you have interaction when you flick the switch and the light comes on, but there is no intrinsic relationship between the turning of the switch and the light. It only becomes interesting when there is an intrinsic relationship between your actions as a user, or as part of the work, and the experience you get in return.

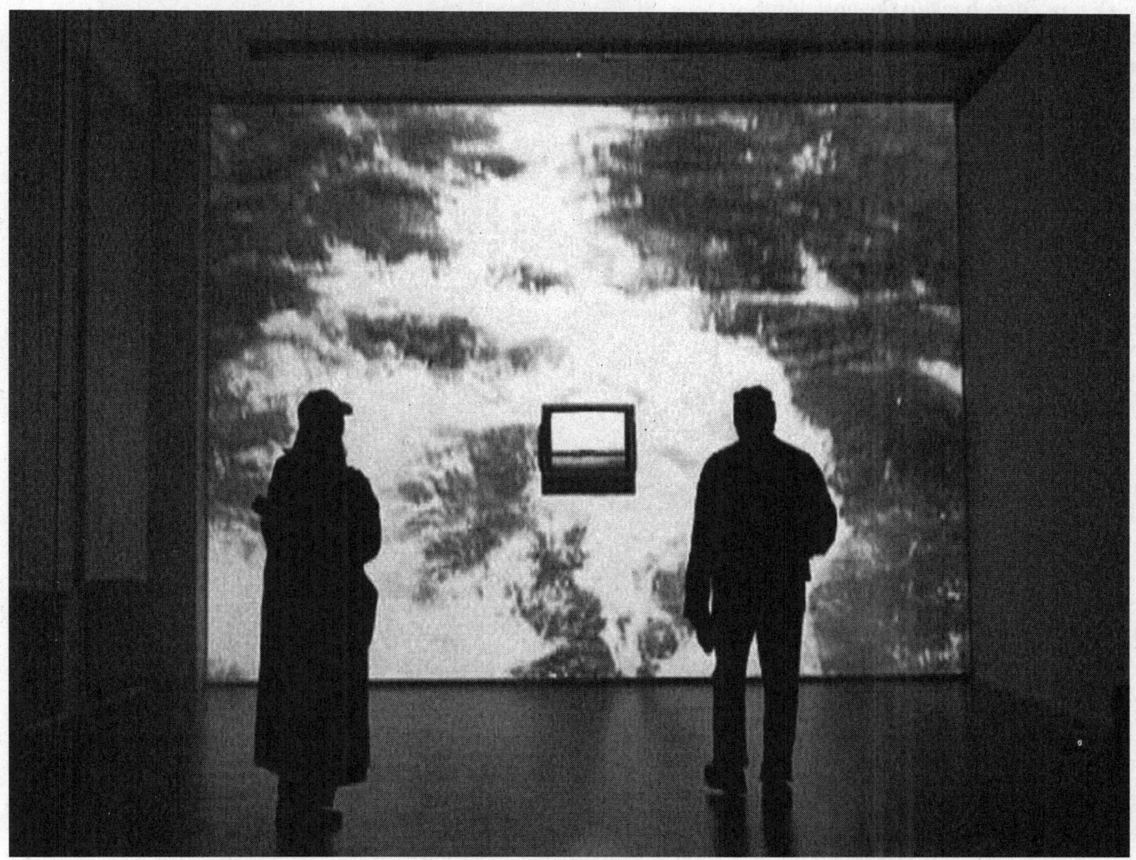

6.29 Geert Mul, *This Land is Man-Made*, installation view, 2000

Anna Tilroe: The Better World

An exhibition as 'the ultimate shopping experience': it was bound to happen. Now that Prada and Gucci, Van Beirendonck and Viktor & Rolf have surrounded their shoes, handbags and clothing with the language and symbols of art, art has no choice but to reciprocate. De Vleeshal's exhibition does just that. As is clear from the title, *Higher Truth No. 5*, this is not about the high street; it's about fashion and branding. Only in fashion can a Higher Truth become a commodity.

This resounds throughout the press release. Vleeshal director Rutger Wolfson and fashion expert Guus Beumer list a series of words that leave you speechless: '*Obsession*, *Higher*, Escape, Freedom, Envy, *Truth*, Eternity, Zen, Egoiste, Addict'. One line after the other, the words in changing sequences, *Higher* and *Truth* stressed every time around, like a mantra. In this case, however, the spiritual perception has a contemporary quality. These elevated notions represent the most fleeting form of luxury: expensive perfume. And so you 'read' them as brand names: *Higher*, Christian Dior; *Truth*, Calvin Klein; and the title's *No. 5*, Chanel.

Already, there are those claiming copyright on words they have given a meaning of their own, a conceptual content that can be used in an argument as if it were a little ©-labelled package. To insiders, lines like 'Techno babes frolic in the techno meadow' require no further definition: that's that.

It's the same in advertising, particularly in fashion advertising. Here too, we practise a form of double reading, taking in several levels of information at once. Thus, something like a D&G logo suffices to add a special dimension to a picture of a groggy-eyed, emaciated woman in striking clothes. It doesn't matter who she is or what has happened to her: fashion is not about reality; it's about mood. In this case, a mood encompassing decadence, chic, glamour and loneliness, wealth and unquenchable desire. Enough, in short, to write a book about. Or to pop by the shop and buy this interesting, spiritual outfit. All this thanks to the brand, attached to the tableau as if it were a label. Without branding, the clothes of Dolce & Gabbana, Comme des Garçons and Armani would never have become the world in itself they are now.

The organizers of *Higher Truth No. 5* have named this world apart 'the parallel universe'. This, they write, is their focus because 'the powers of imagination employed in conjuring up such worlds is perhaps the most fascinating aspect of fashion. Advertisements, shops, fashion shows, models – all serve to create an entirely visually defined, more beautiful, more stylish, *better* world. Yet when fashion is shown in a museum setting, this imaginative power is passed over. Fashion is always shown in the form of a garment.' This is a convention they want to break away from – and so not a single dress or pair of trousers has been put on display in De Vleeshal.

They have a point. Things are, indeed, brewing in museum country. The policy of museums to focus exclusively on the aesthetic qualities of objects, and ignoring the context in which they originated, is under fire. Museums, particularly museums of contemporary art, do not hold with describing the social and political climate in which art is produced. They prefer to focus on the art history connections that can elevate a piece of art to the status of timeless relic. As a result, art forms highlighting everyday life or not bound to objects, such as the performance of the 1960s, socio-political art and present-day VJ art, hardly feature in collections – if at all. And so they are awarded no significance in terms of art history.

Such criticism is nothing new. Artists turned their backs on museums in great numbers as early as the 1960s; political pressure was brought to bear as well. It was hushed up during the 1980s; today, once more, there are those who claim that the fuss will blow over. I doubt it. Now that imagination is no longer the exclusive province of the arts, but an essential part of virtual reality, the market economy and its resulting social developments, museums are more than ever being asked to take a stand. How does 'the house of art' respond to a world in which reality and the unreal, imagination and strategy are becoming ever more intertwined?

At De Vleeshal we see an exhibition that is not an exhibition, but rather a magnified reflection of the excessive way in which we have lost touch with reality. Here the 'parallel universe' turns out to be a shop that consists purely and simply of design. Every single detail – the entrance, the counter, the changing rooms, the clothes worn by the staff – has been conceived by a group of designers, architects and fashion designers (Herman Verkerk, Rianne Makkink, Jop van Bennekom, Project Jordan, Alexander van Slobbe). In so doing, they have kept the principles of the attention economy firmly in sight. Those principles have been laid down in one of the two hefty volumes of the recently published *Harvard Guide to Shopping*. The book is the report on a study of all aspects of shopping, carried out by Harvard students and supervised by Rem Koolhaas. It shows the high-end design shop to be composed of four zones, each with its own mood. The zones reflect and stimulate the psychological processes customers go through from the moment they enter the shop to the moment they pull out their creditcard.

Zone 1 of De Vleeshal, the entrance, is a small hall with glimmering brown and orange pillars, which serves as a waiting room. Here, we (the customers) can recover from our 'street experience' and prepare for what is to come. The beautiful young man dressed in designer clothes by Alexander van Slobbe is a big help. Bowing slightly, he opens the large door in the austerely designed aluminium and copper coloured dividing wall with a flourish. We enter Zone 2. This is the heart of the shop. It's surprisingly bare. Only the two long walls have been fitted with displays and a counter. But of course: a crammed shop is not chic. That is for consumer society, for the average Joe. The high-end shop is the equivalent of *nouvelle cuisine*: each pea on its own plate. Once again: beautiful, well-dressed staff; apparently intelligent too. Abstract techno music and gorgeous lighting making the walls look alternately transparent and mat. Everything as cool and clear as the water in the water feature.

Lovely. So why are we reminded of an essay written by Slovenian philosopher Slavoj Žižek in response to 9-11? 'The distinguishing feature of the 20th century,' he writes, 'was the passion to penetrate through the web of pretence out of which we have constructed our reality, to what is Real. However,' he grumbles, 'in this day and age we can only distinguish what is Real from pretence by staging it in an artificial spectacle. The hijackers grasped this fully [...] What we now experience in the capitalist, utilitarian, secular universe is the *dematerializing* of "real life".'

Gloomy words; we try to shake them off as we walk towards a door at the end of the carefully staged emptiness. It takes us to another waiting room, Zone 3, by way of a narrow passage. The floor, which is now noticeably carpeted, slopes here, and continues to do so all the way through the tubular corridor leading to the highest point of the shop, Zone 4. The changing room. Here we are alone in a way befitting the times: a large monitor shows us entering and looking around in slow motion. Much like psychologists observing purchasing behaviour, we are mesmerized. But what do we see? Are those our own eyes watching, or are we looking at ourselves through someone else's eyes? Does our view of ourselves match the shadow hot on our heels?

The sense of emptiness has now reached its peak: there's nothing there and yet we have experienced something. Something executed with the same meticulous sense of detail as a Zen garden. In four carefully set up stages we have discovered the parallel universe to be nothing more than our own carefully styled, unquenchable desire for the next peak experience. This desire for the 'perpetually new' is our society's driving force. It is the source of its dynamics. Fashion's role is to nurture and style that desire.

Nobody can escape from fashion. Its ubiquitous pressure is even felt by those who couldn't care less about the latest car models, photo cameras and bikini lines: they are singled out as old-fashioned, conservative – even as misfits. The unfashionable do not partake in the dynamics of the times, in the whirl of the ever-new. They miss out on the feeling of excess resulting from a fashionable purchase, the kick of acquiring a new look, identity or status, and the pride in owning something that brings them up to date. And what, in this media-governed age, is more sensational than the 'in thing'?

'An artificial spectacle,' Žižek grumbles once more. Yet he, too, knows that a powerful force lies hidden behind pretence; what's more, that we need pretence in order to gain some understanding of what reality actually is. Art is pretence. Art fleshes out that which we suspect but don't really know with imagination, in hopes of catching a glimpse of what is real, pure and authentic. Without imagination, we would wander desperately through a vast emptiness.

This has also caught on beyond the domain of the arts. Branding is an immense economic process revolving around the phases created by imagination. To these, idealistic notions are often attached. And so Rem Koolhaas, the architect now also interested in fashion, states in the book he produced for Prada: 'to add intelligence to an object [...] is the only way to advance [...]. Intelligence is an altruistic gesture that increases the value of an object.' In other words, the world will become richer and mankind will change for the better if we project our ideas imaginatively onto objects. Which raises the question of what we should do with those objects. Replace them with other intelligent objects? Again and again, as dictated by fashion?

Feeling dejected, we leave De Vleeshal's changing room and make our way to the counter in Zone 2. Now that we have a better understanding of the parallel universe, the beautiful people encourage us to buy a ticket. We show it to the doorman at the entrance with a sense of relief. And out we go, out into the unintelligent chaos of the banal common-or-garden-variety world.

Guus Beumer: The Language of Fashion

1. Fashion as an Anachronism and as a Perspective for the Future

Over the last few years, fashion has acquired a strange kind of allure. Strange, because the fascination has not been confined to a specific group: almost everything and everyone has been enthralled. All of a sudden, artists, designers and architects prove themselves willing to respond to a discipline that, only a decade ago, was dismissed as a vain exercise in futility. Hundreds of articles have been published singing the praises of the symbolic, psychological, semiotic and, of course, economic qualities of fashion. What all these articles have in common is their implicit respect, their awe, for a discipline that seems to trip gracefully along the surface. Not because of this skirt length, or that supermodel, but because, parallel to the growing dominance of market forces, fashion has developed into a system that dovetails perfectly with the dance of supply and demand and so succeeds in seducing a global market.

How could fashion, of all things, have found the answer so many other disciplines are still searching for? How has an anachronism like fashion managed to mesh seamlessly with the perpetual here and now of the market place? For it *is* an anachronism. Whilst almost every other discipline has lost its faith in renewal, fashion has elevated this belief to its *raison d'être*, dedicating spectacular services, in the form of hundreds of shows, to 'the new' twice yearly. But this is just a detail, a sideline in the labyrinth of fashion.

How and by what means does fashion correspond to current trends in the market place, so that it becomes a perspective for the future? For that is what fashion truly is: there is not a single discipline that has not, although sometimes timidly, looked with envy upon fashion's seductive powers. Understandably, for now that the hierarchy differentiating high and low has made way for the hierarchy differentiating fast and slow, the dynamics of fashion, as well as its allure, have become exceptionally desirable.

This article aspires to uncover the language of fashion and to show how the premise of this language contains the paradox set out above. It subsequently aims to show how this language has become the language of consumers, to conclude by describing how this language is changing the classical discourse on culture. To do this, we must broaden our outlook. Away from the designer, away from the catwalk or the shop and towards the industry: that powerful conglomerate of capital, production and distribution that lies hidden behind fashion's allure.

2. From Elite Sophistication to Mass Consumption

It must have been in the last century, at some time in the late 1940s, that fashion began to take on its present shape. The Second World War had destroyed almost all of Europe's traditional infrastructure, and the United States set itself the task of building a modern industrial system atop the ruins. A clothing industry, as well, but in order to do that the American advisers first consulted their own chemical industry, the DuPonts of this world. For it was American chemistry that provided the perfect solution for Europe's devastated textile industry of natural fabrics like wool or cotton: synthetic fibres. To the Americans, synthetic fibres were the future. They believed these to be more practical, more durable and, as they came from the oil industry, they had the added bonus of substantial profit margins. What was lacking was effective marketing; for that, the advisors looked back to the old continent, to the couturier.

The couturier was a quintessentially French invention, the product of a hyper-refined court culture as could evolve only in France or, to be exact, in Paris. To be sure, this culture has disappeared in its original form, but, in one of history's stranger turns the French Revolution preserved, rather that destroyed, various aspects of it. The hierarchical organization of French bureaucracy is a case in point, and here the concepts of an elite culture, and thus of couture, continued to be of great importance long after the revolution. To new American money the notion of aristocratic Paris, of a cultural elite, was irresistible. The scions of this class of newly rich were the ones to visit couture's salons. After all, to invest in couture was not just to invest in a new dress, but also in luxury, in sophistication, in an identity and all that aristocratic allure. But that was all before World War II, and post-war American industry sought a strategy to remodel this desire for luxury, for identity, and so to make it available to greater numbers of consumers. How, in other words, could couture, the elitist world of one-offs, be harnessed by the industrial apparatus?

To the American marketers, the solution was simple: make use of the couturier's information and transform that unique image into a mass product. Yet couturiers refused to co-operate and resisted the trivialization of their visionary powers. A bizarre contest enfolded between couture and industry, between elite sophistication and mass consumption, between Europe and the United States; the prize was press admission to the couturier's chic salon presentations and so to information.

For couture – in its most classical form – did not communicate with the press, but with individual clients. The press was merely an onlooker, representing a public wanting a taste of the elite's dream. However, industry demands transformed that same press. It came to represent the industry rather than the public, and as such, initially met with couture's opposition. Eyewitness accounts tell the story of this early struggle. The couturier's resistance was relentless and persistent. Cameras were kept out, handbags were searched, every pencil stub and scrap of paper was banned from the salon. The press struck back. Fashion journalism was never as expressive, as metaphorical as it was in the late 1950s. The industry gratefully seized upon the florid descriptions. Language as a means to evoke images. Couturiers made another countermove: no longer describing their unique creations as 'svelte, sleek *demirobe* in grey crêpe georgette', they only gave numbers – no. 27, for instance – to their brainchildren. But, again, they were unsuccessful; ultimately, as is always the case when the wishes of tradition clash with the demands of industry, they were to be worsted. Not necessarily to their detriment – quite the opposite. As a brand, the couturier would achieve immortality. And those same mistrusted media would endlessly elevate their status, so that the global consumer would also come to believe in the elite's quest for identity. Moreover, both the couturier and part of the press would metamorphose: the couturier into the prêt-à-porter designer and the fashion journalist into the stylist.

3. The Triumphal March of a By-product: Styling

According to the official history of fashion, the transformation of the couturier into the designer of ready-to-wear, of prêt-à-porter, was of decisive importance to the future of the discipline. The unique design had made way for the mass product, signalling the industrialization of fashion. However important this change may have been, it is overshadowed by a second transformation: that of the fashion journalist into the stylist. For it was this turn of events that would eventually cause fashion to break away from its classical product – clothing – providing a new perspective that would enable fashion to dominate practically every other discipline.

Pierre Cardin, Courrèges and of, course, Yves Saint Laurent belonged to the first generation of couturiers to present prêt-à-porter collections in

addition to their series of one-offs. This transformation is considered revolutionary – revolutionary
in the sense that the new lines were industrially
produced, and therefore much cheaper than
the couturier's unique – handmade – creations.
Apart from economic considerations, price is also
important in establishing image. Lower prices made
it possible for fashion to rid itself of its elitist image,
for it to claim the title 'democratic'. For democratic
was a keyword in the turbulent 1960s! Thus prêt-à-
porter, the invention of American industrialists, not
only went down in history as the French answer to
Britain's budding pop culture, but also as the onset
of the democratization of fashion. Freed from the
salon, from the tentacles of the elite, democratic
fashion would finally be able to respond to the
streets. But in order to stay current, another change
quietly took place within fashion, one much more
important than different prices or production
methods. Fashion would no longer, as it did in the
days of couture, express personal statements about
what was new in terms of the present, of the here
and now; it would make more general pronounce-
ments about the new as … the future!

It is this statement that permitted a new
fashion caste, a new product and thus a new
language to evolve. The introduction of prêt-à-
porter also gave birth to the stylist, who, using a

strange hodgepodge of image and text, revealed what others could not see. The stylist's product has many names and is also known as *mood*, *prediction* or *tendency*; however, the most popular term for the new product is *trend*.

But why the future? Why was it necessary for prêt-à-porter to introduce the future? Fashion is an industry and prêt-à-porter is industry's invention – and this is where the answer lies. The introduction of the future was necessary because of production schedules. In the age of couture, producers had the problem of what is called *lead time*. By the time a new 'look' appeared in the shops, the couturier had already presented the next collection. Giving glimpses of what the future might bring provided the industry with a response to the dynamics of fashion. And with the opportunity of getting the product to the shops on time, meeting the requirement that the industry stay current. Moreover, it gave the industry's ally – the press – ample time to inform the public about what would be 'current' in fashion.

This explains the paradox of fashion, a discipline that appears to be at once an anachronism and a perspective for the future. Fashion's fixation on the future is not ingrained; it is no remnant of a distant past, no echo of modernism. It has in fact only recently been introduced, somewhere in the early 1960s, for purely pragmatic reasons. And so it seems that the paradox of fashion can only and simply be understood as a result of its industrialization. We, the consumers, are the true anachronism – in our abiding readiness to believe in this projection of the new as the future. Fashion merely reflects this willingness, for as far as fashion is concerned, consumers hold the truth.

4. From Text to Image, from Stylist to Trend Watcher

Looking back, it is the speed of the fashion journalist's transformation into the stylist that is most remarkable. As a consequence of fashion's many style changes and their attendant high financial risk, the new profession is increasingly called upon. When dealing with the top echelon of the market, the designer is still an essential part of the marketing mix, but democratic fashion can – indeed *must* – address the market as a whole. In this respect, the designer's vision is not simply too specific, too personal; it also offers the industry too little security. The industry has to anticipate sooner, faster, better, in order to develop not only the image of fashion, but also the required threads, fabrics and colours, whereby the question is not so much one of looking, but of working ahead. Just as it is the designer's job to literally shape what is to come, the stylist has the task of conjuring up a crafty self-fulfilling prophecy by suggesting a prediction (of what is to come).

The medium used by the stylist has, over the years, evolved into a unique mix of text and image. In the early years, however, the print journalist's traditional background predominated, and so text reigned supreme. Literally thousands of letters were sent from Paris and London to the United States. Letters about favoured silhouettes, popular fabrics, the right colour; letters about details such as button fastenings, pocket flaps, collars; letters about the way lipstick was applied, hair dressed in coils, about the decoration of the salons – every detail was important, for everything held the potential for change. It did not take long for a publisher to decide to bring all that information together, and so the American trade magazine *Women's Wear Daily* – still trendsetting today – was born. The advent of this periodical was the zenith of the stylist as a gatherer of information. From this moment on, the journalistic background and the status of text as the preferred medium would recede into the background, and stylists would undergo their very own development. To trace every single transformation of this profession would take us too far afield; what is important is how the stylist eventually, sometime in the late 1970s, became a trend watcher.

Ah, the trend watcher. Is it not a lovely, mystifying term – and in that sense not a fine, industrial response to the designer? A response that the more pragmatic term stylist cannot provide, being purposefully stripped of the aura of designer/artist. But, as it turned out, aura also proved to be an essential ingredient within the context of industry; it was reintroduced by bringing in the term trend watcher. A term denoting not as much a profession as an invention and which, of course, must be taken as a harbinger of a world in which everything was to be dubbed as ephemeral or, better yet, virtual. 'My product is a service in the form of information,' says the trend watcher, and this statement marks the advent of this new language. What kind of language is this? Or, to put it another way, what does this language look like, as it is not so much a spoken as a visual language?

Instead of the copious descriptions produced by the journalist/stylist, 'trend books' with 'mood boards' came to be the trend watcher's main tools. From this moment on, mood boards – or collages of images – of water, for instance, with a single notion, such as 'purity', bringing together all their diverse elements, would not only reveal a future, but also colour ranges, preferred materials, a visual language and a new outlook.

Remarkably, notions like purity are used as images rather than words, their foremost function being to give thematic direction to the diversity of the collages without diminishing their ambiguity. It is precisely this ambiguity that enables the trend watcher to be both vague and specific and to simultaneously surpass all disciplines. For it is in this guise that the trend watcher discloses not so much tomorrow's clothing as the spirit of the times. And the spirit of the times does not control only the clothing industry; it also affects interior design, food, the choice of new technological games, body weight and the types of holidays taken. Or at least this is what the market would like consumers to believe, so that the fragmented consumer's chaotic behaviour can still be anticipated.

5. The New as Atmosphere

Behold the trend watcher's multi-disciplinary language: stripped of history, of craft, of discourse, addressing everyone and thus accessible to all. Yet the trend watcher's language is limited. It is the language of the pragmatist who understands that simplicity – water is transparent, transparent is pure, pure is Zen – communicates. In the last few years, for instance, the notions of minimalist, modernist, classical and Zen have been the order of the day; a series to which Baroque has now been added. The trend watcher's job is to chart tomorrow's world, using this handful of catchwords. Some truly remarkable combinations, like minimalist classical, minimalist Zen, modernistic classical, minimalist modernist, Zen modernist, Zen Baroque, minimalist Baroque, et cetera, have been the result. However directional these combinations may be within the visual wealth of a possible future, they also reveal the limitations placed upon the trend watcher. Trends are never explained; trends are only indicated.

In the end, the trend watcher's thematic trend books reveal as little about the future as a fortune-teller's crystal ball. Still, an important transformation has taken place. The future has had to pay a price for its would-be visibility: it has been stripped of all context, ideology and perspective. Today, newness is merely atmospheric.

Limited or not, this new language does communicate. The trend watcher's language has become the official language of many a fashion and interior design magazine. And so, too, of many a consumer, for nowadays everyone is perfectly able to follow and use the thematic, trendy language of the trend watcher. Now, neither minimalism nor modernism holds any mystery for modern consumers. Stripped of their historic and ideological content, both notions have become synonymous with an interchangeable aestheticism that can be applied to any discipline. Minimalist Zen modernism, for example, stands for graphic-like divisions of surfaces and three colour groups (grey, white and brown). The difference lies mainly in the medium, for, thanks to trends, even the problem of disparity in materials has been overcome: a collective preference for felted wool commands both the clothing and furniture industries. Which proves that the language of fashion has broken free of the limits of the clothing industry and that several disciplines have become subjected to the trend watcher's style principles. An observation that is further borne out by examining the differences between the classical interior design magazine *World of Interiors* and the relative newcomer *Wallpaper*.

6. *World of Interiors* versus *Wallpaper*: the Unique versus the Trend

World of Interiors was the pre-eminent interior design magazine of the 1980s. Fat, shiny and snobby, the magazine is an ode to the individual and the gift of creating a unique living environment. It does not merely focus on designers' decorative talents; its spreads also endeavour to provide a almost three-dimensional experience of the exquisite examples of (interior) design. Designers are seldom portrayed. They remain anonymous, present only in these tranquil interiors. The same applies to the users: they, too, are out of the picture. This world of interiors revolves around itself, deriving its *raison d'être* from its unique aestheticism.

Wallpaper, on the other hand, first published in the late 1990s, is not an interior design magazine in the usual sense of the word. Magazines like *Elle* still feature clothing, interior decoration and food under different headings, but *Wallpaper* mixes the different disciplines into an single exercise in style. Taking its cue from the trend watcher and on the basis of his language, *Wallpaper* proclaims fashion not only to be clothing, but also food, gardening, cars, holiday destinations, the workplace, preferred sports and the outfits worn playing them. Whilst *World of Interiors* approaches interior design as a separate universe, *Wallpaper* relegates everything to the realm of fashion, aestheticizing every aspect of life. Employing the trend watcher's thematic notions, it is still possible to present the consumer's fragmented world as an entity … but, this, too, is not without its consequences.

Wallpaper is bound by the same restrictions as the trend watcher. Removed from any kind of context, *Wallpaper*'s world has a strange neutrality. Other British magazines clearly have social roots; *Wallpaper* is classless, apparently springing from a vacuum. There is not the slightest hint of *Tattler*'s snobbism or *I-D*'s taste for the authenticity of the streets. *Wallpaper* is as transparent as the crystal-clear aestheticism of minimalist Zen modernism. This artificiality is also evident in the photography. While *World of Interiors* presents unique but real interiors, *Wallpaper* focuses on the trendy interior, often created expressly for the purpose of the photograph. In *Wallpaper,* space is two-dimensional. The rolling spreads of *World of Interiors* are replaced by an abundance of medium shots and close-ups that have no other purpose than to illustrate the thematic relationship between the décor's different elements.

Still, unlike *World of Interiors, Wallpaper*'s pages are inhabited. In almost every series, people are put on stage with the food or furniture. This does not make the settings any more real. On the contrary: the food, clothes, furniture and people serve – at best – only as representations of a particular trend. In *Wallpaper,* the theme does not just serve as the backdrop; it is all-important, the main character of the show: Zen model in Zen clothes eating a Zen snack on a Zen couch.

7. The Language of the Trend Watcher as the Language of Architecture

How has the trend watcher's language influenced disciplines other than that of design? What influence has the trend watcher had beyond the field of not only fashion, but also the realm of lifestyle? Where has this seductive force left its mark?

These traces are readily apparent in architecture. Take the present (postmodern) fixation on the surface as a building's canvass and the necessity of furnishing this 'skin' with new materials, over and over again. Or the unstoppable rise of the stylist in the guise of the interior designer, with results such as Philippe Starck supplying ready-made dwellings, assembled as dictated by the latest trends. The image of the trend watcher's influence spreading like an oil stain, its reach extending further and further – from clothes, to interiors, to a building's skin – is impossible to avoid.

The trend watcher's influence, however, reaches even further. It is not just the trend watcher's product, but also and especially the trend watcher's language that is influential. So much so that when it comes to architecture – the mother of all arts – the question arises whether a new mix has not replaced classical discourse.

8. Of Image as Language and Language as Image

It is remarkable that the present success of Dutch architecture is ascribed solely to its quality, to the talent that just happened to emerge in the Netherlands. The strategic aspect of this success is either glossed over, or neutralized by pointing out the stimulus the authorities have provided by commissioning works. Yet it is precisely strategy that must often have been a decisive factor.

This is evident from the fact, for instance, that many Dutch architects have produced publications highlighting their work in recent years. *SMLXL*, by Rem Koolhaas/OMA/Mau, has undoubtedly contributed to this – an opinion commonly held, but one generally based on appreciation for this Bible among books on architecture. *SMLXL*'s importance as an instrument of strategy is ignored. Yet this is where the book is most innovative, beginning with the organization of the information: on the basis of the notion of 'scale' and 'objectively', alphabetically – resulting in an unusual approach. As though the arrangement according to scale tells the main story, with the alphabetical lay-out – like a

Greek chorus –proving any organization, including one into Small, Medium, Large and XLarge, to be ultimately futile.

This dichotomy can be perceived as intellectual shadow boxing, although this entails the risk of rendering all knowledge insignificant: stripped of any clear perspective, no more than an entry in an encyclopaedia – merely a stepping stone to the next piece of information.

Equally important are Bruce Mau and the way he interconnects text and image. Imagery has an illustrative role in classical architectural discourse, as a mere accompaniment to the text. Mau assigns equal importance to imagery and text, or even has text take on the role of the image. Moreover, in *SMLXL* any image – even an advertisement for underpants – proves usable, and Mau has abandoned the classical representation of architecture – that of the building as an object. It seems the trend watcher's mood boards have not been lost on Mau.

The result of this collaboration between OMA, Rem Koolhaas and Mau is inescapable and can lead to only one conclusion: *SMLXL*'s universe is round, complete and true, but this, too, has its price. By so expressly creating its own context for a body of work, *SMLXL* has removed itself from classical debate, and all information – text, street images, advertisements – is reduced to illustrative material for this universe.

SMLXL is a historic event. Undoubtedly, this is the book that 'launched a thousand ships', but not primarily because of the book's content, but because of the strategic fact that, in *SMLXL*, the discourse has become the projection of an identity.

The success of *SMLXL* quickly found a following, and MVRDV and UN Studio published equally fat books on their collected works. But whilst Koolhaas gives a slanted and thus unique anthropological perspective to his work, these agencies mark out classical architectural and philosophical themes such as 'density' and 'fold out' as their exclusive territory. This reduces architectural debate to an exchange of single notions which then proceed to function in much the same way as do fashion's logos. Successfully so, for MVRDV has by now become synonymous with density. The same will undoubtedly happen to UN Studio and fold, once the recently published book *UN Studio-UN Fold* has reached the general public.

This evolution has not yet reached its conclusion.

For the past few years, Koolhaas has been steadily building his own conceptual framework that should enable him to interpret new phenomena. One example is the word 'bigness', which he uses to denote a recent phenomenon exceeding the present frame of reference of human size. He could never have used a word like 'megalomaniac' for this phenomenon, as this word has too many political connotations. Here, bypassing the existing conceptual framework and so neutralizing excessively straightforward meanings proves productive. Yet some caution is called for. It is not difficult to understand why, in the interests of visibility, a concept such as bigness should be introduced, but does the same go for notions like 'chiness' or 'scape'?[1] Is visibility the primary motive here, or is language being artificially created to function with the same kind of directness as a logo or an image? Such neologisms may have communicative powers, but do they do anything to further the cause of visibility? The world described in terms of bigness, chiness and scape? A language like the trend watcher's, stripped of history and context?

What's more, Koolhaas has taken steps to 'brand' the concepts he has created, like chiness and scape, as his own. Why? It may, theoretically, be possible to claim copyright on neologisms,

[1] Chiness: a Chinese identity; scape: neither city, nor landscape – both neologisms coined by Rem Koolhaas, from the Pearl River Delta project.

but this does not give Koolhaas the right to appropriate the meanings of such notions, or the phenomena the words are supposed to describe. Here we have the age-old problem of form and content: form, in this case, being something that can be protected, but content permanently part of the public domain. Why has Koolhaas nevertheless sent concepts into the world under copyright notice – even though the meaning he attaches to that copyright notice has lost most of its legal validity? One possible explanation is that Koolhaas is launching these concepts and their copyright notice as signs, as some kind of registered trade-marks™, much like logos, and that he is attempting to appropriate these signs, using the suggestion of trademark legislation, in the process trying to delay the inevitable annexation of these notions and images by other interested parties.

Which would also prove that Koolhaas is purposely putting his vision forward as a marketing strategy, intended to strengthen his position in the face of the competition: other architects.

One thing is clear; in his quest for visibility, the trend watcher has developed a language that can neutralize a notion, a phenomenon, a development, by removing it from its context. This language, which has taken on the immediacy of an image and in which everything can be presented as new, as a potential, is no longer exclusive to fashion, but has become part of the language of discourse. And the annexation of these neutralized concepts – as images, as neologisms, under copyright notice, as logos – is no more than a logical step in a culture dominated by market forces.

9. A Moral Question

It is time for a moral question. What is the significance of the transformation of thinking from a open place from where a future could be created to a closed universe in which a future can only be represented? What is the significance of the evolution of the discourse into a market tool and so into an instrument of power?

Surprisingly, this question can be answered in much the same way as it was a few decades ago. The link between discourse and power is nothing new. The thinkers belonging to the Frankfurt School demonstrated power to be embedded within the discourse itself. This power, the Frankfurters believed, had to be acknowledged in order to be neutralized. To the Frankfurters, a discourse free from power is both a means and an end.

The Frankfurt School is now regarded as an after-effect of Marxism and dismissed as one of history's mistakes. Still, it can be interesting to take a look at the present day through the eyes of the Frankfurt philosophers. Seen through their somewhat red-coloured glasses, the evolution described above, the appropriation of parts of the public debate as the exclusive domain of an architect or a designer, is easily interpreted as a variant on the struggle over the means of production. Seen from the Frankfurter's perspective, the present debate on architecture is characterized by the struggle for the power to attach meanings to notions and to subsequently claim them as identifying concepts. The question then raised is: how to liberate this evolution from its power structure?

But *should* this evolution really jettison its power structure? Should the question not be considered to be out of date, as are the Frankfurters? To what extent do moral outrage and Marxist struggle form an appropriate response to this social evolution? After all, the discourse has proved to be a highly profitable marketing tool. Dutch architecture, with Rem Koolhaas as its figurehead, is internationally successful and has become a model for cultural enterprise. The portfolios of OMA, MVRDV, UN Studio and

many other agencies are filled with commissions,
as opposed to the 1970s and '80s, when architecture appeared to have been reduced to a 'paper
art'. Would it not be legitimate to ask whether the
transformation of the discourse into a marketing
tool has not simply heralded a new phase in the
discourse, reflecting present social conditions? Has
the historical position of academic discourse not
inevitably made way for a more current exposition,
corresponding to the requirements of the market
place – thanks to Rem Koolhaas *et al.*? Or does this
all sound too pragmatic, even unscrupulous?

10. Starting Over

More interesting, aside from potential criticism,
moral outrage and war cries, is that this same architecture is already making attempts to escape from
the evolution described above and its attendant
power mechanisms.

In 2002, under the banner of 'Starting Over',
Forum Magazine recently published a *Weekend
Agenda*: the outcome of four weekends of retreat.
At the invitation of *Forum Magazine*, a group of
journalists, artists, (fashion) designers and architects withdrew to the country. Leaving their own
disciplines, their particular mediums, discourse
as strategy and strategy as discourse behind them,

they were instructed to rework a heap of rubbish into a joint construct. Their goals were to achieve new relationships with each other and with the rigid inflexibility of their materials. What came first to this group was not so much the act of building, but their ambition to create a parallel universe, separate from the prevailing realities of the marketplace.

An ambition that has hitherto gone largely unnoticed. The only one to write about the agenda was Hans Ibelings, in *de Volkskrant*, qualifying it as 'meaningless', superfluous criticism. For in this time of 'bigness' all such ambitions are meaningless. What is important is that an attempt was made anyway, in full awareness of public debate being exposed to the same forces as is the public space.

11. The Trend Watcher's Language and the Discourse on Art

How much influence does the language of fashion have? Has the emergence of the trend watcher had the same far-reaching consequences for art as it has had for architecture? The trend watcher's strong point is the ability to remove everything from its context, to neutralize everything so that each discipline, each product, each style, each perspective, each outlook can be exploited as something new, as the future, as having potential. What influence has the trend watcher had on discourse within modern art? Modern art is a discipline that simply cannot do without discourse. Fashion has no need of debate; it needs consumers. But to art, discourse is vitally important, preserving the image of art as the new metaphysics. Any change in this discourse is of crucial importance to art, and it must have left its marks. Where and how has the trend watcher's influence manifested itself? Has the language of fashion become the language of modern art?

Almost every aspect of modern art is indebted to the marketplace, but the defining influence of supply and demand is mostly clearly present in the production and distribution of art. A painting's size and the time and place of its sale are subject to the interests of the marketplace. Yet, strangely enough, reflection on art seems to have escaped that all-powerful market. How is this possible? Despite price agreements between museums, gallery owners and artists, despite publicity deals with the media, despite the economic dominance of established institutions, the language of art is still more or less the same as it was 10 years ago. How has the discourse on art, as opposed to that on design or architecture, managed to remain so – seemingly – intact, preserving its purely academic, largely philosophical roots? And how have artists, like architects, managed to create an exclusive context for their work, incorporating a particular reflection and criticism into their work – yet without taking parts of the discourse on board? Could it be that the large and powerful institutions in which art is embedded offer protection, somewhat staving off the influence of the marketplace? Is it only a matter of time, or has art really escaped the trend watcher's language?

For the moment, art *has* eluded the trend watcher's language, but the question remains: how? How has the reach of the all-powerful marketplace been limited to the production and distribution of art, ignoring reflection? There can only be one answer: because this is in the market's own interest. Just as the creed of the new is a necessary part of the marketing mix of fashion, academic discourse is inextricably bound up with modern art. Any change, including appropriation by third parties, would damage art, for instance by transforming it into design. Such change would have far-reaching and considerable economic consequences for art and would hardly be in the best interests of the market.

Thus, in the year 2002, the Frankfurters' goal of relative autonomy for (the discourse within) modern art has not been achieved in spite of, but because of market forces. But, again, at what cost? Is the present discourse on art merely a shadow dance put on in the interests of the marketplace – as is fashion's fixation on the new? What truth lies behind the truth of the present discourse? What becomes visible and why? Whatever the case, the conclusion formulated above renders the dichotomy between commercial and non-commercial art fictitious. For the discourse itself is an immanent part of the marketplace.

It may be 'meaningless' or naïve, but the Documenta 11 (2002) in Kassel just happened to have 'Starting Over' as its motto – just like the *Weekend Agenda*. Not only was that other dichotomy – the ethnocentric distinction between North and South – placed on the agenda, but also the relationships currently prevalent in the market place. It is interesting that such questions are raised not only by individual artists, but, as Kassel shows, also on a larger scale, in one-off projects and within a multidisciplinary framework. The relative autonomy of art, realized – oh, irony – by market forces, clearly facilitates the formation of such interdisciplinary alliances in the art world. These alliances could produce a new dialect that might eventually transform the present dialogue, crossing the current divide between production and reflection in art. The question then arises: what will uncover this Esperanto, fostered not by philosophy, not by the trend watcher, but (for instance) by the activist, the anthropologist, the scientist or the politician?

And what about the trend watcher? The trend watcher has now gotten wind of a desire for something apart from the marketplace. The theme: Romanticism!

With thanks to Marcel de Zwaan

Chris Darke: Sublime Bodies. On the Work of Chris Cunningham

'I'm freaked out by the body. I used to pretend that I was hollow, because I couldn't stand the idea of having all those organs inside; it made me feel sick.'[1]

A few years ago I was making magazine items for the British TV channel Film Four, and one of the areas they were keen we cover was music video. Inevitably, the name of Chris Cunningham came up. The director had already made a number of striking videos for left-field recording artists such as Squarepusher, Portishead and Aphex Twin, and I remember watching his showreel with the production team, all of whom seemed to be perfectly at ease with the TV priorities of being 'cool', 'contemporary' and 'cutting edge'. I had my doubts. Yes, the work was as 'cool' as the artists whose music it illustrated, but something in me resisted being thoroughly impressed. I distrusted the production values, the high-tech SFX sheen, the shock effect that Cunningham was associated with, but in a way that had as much to do with my own prejudices towards the music-video format as with any real critical discrimination.

A filmed interview was arranged with the twenty-something director, and we headed off to

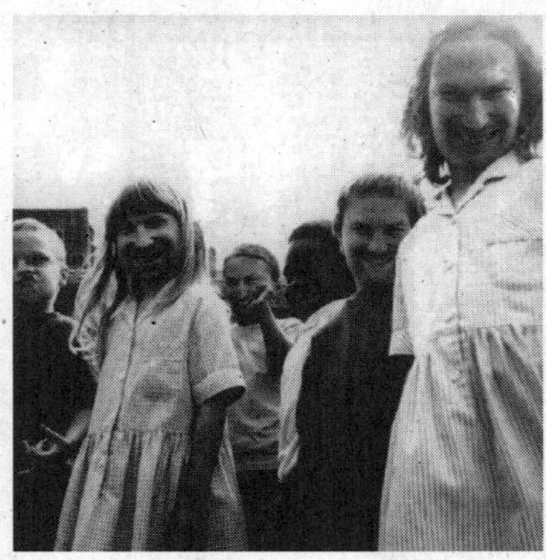

7.1   Chris Cunningham, *Come to Daddy*, album cover artwork, 1997

[1] Theodore Roszak, *Flicker*, Bantam Books, London, 1992, p. 440.

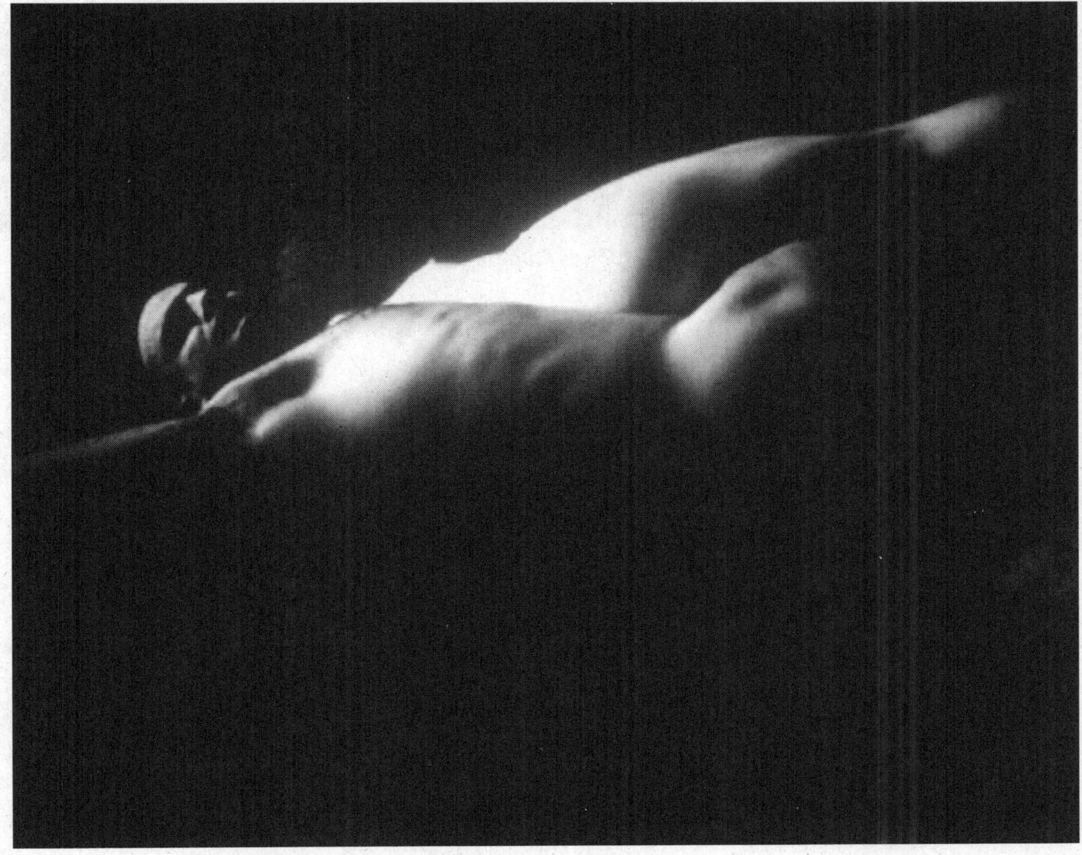

7.2   Chris Cunningham, *flex*, film still, 2001

the production offices where Cunningham was based. In order to visually enliven the standard 'talking-head/video-clip' format of the piece we were filming, the crew assembled a makeshift set out of the materials to hand. Fashion mannequin body parts lay scattered about in the basement room. The cameraman drew on his art-school training, and a Hans Belmer-like arrangement of lifeless limbs and dead-eyed heads was soon installed around the hot seat. Artfully lit, this *mise-en-scène* made for a pleasing establishing shot. Enter Chris Cunningham. Promo-video wunderkind as grunge diva. 'Doesn't like interviews', I'd been warned. Who does? Least of all the interviewer when confronted with a deliberately monosyllabic film-maker squirming in a chair, picking uncomfortably at fraying jeans, his eyes obscured by a lank fringe. Cunningham's body language said it all: 'Don't ask.' I remember the interview as a slow process of coaxing sound bites, like getting a recalcitrant teenager to come clean. And afterwards I promised myself that should I ever have to write about Cunningham I'd open with a malicious literary reference. For me, the film maker was the embodiment of a character from Theodor Roszak's cult novel *Flicker*, in which an old-school humanist *cinéphile*, Jonathan Gates, comes across the work of a young film maker, Simon Dunkle. A stuttering, antisocial albino, Dunkle is a brilliantly gifted tyro whose

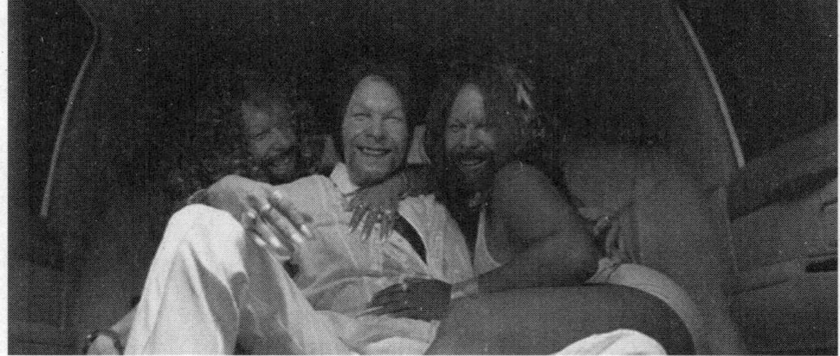

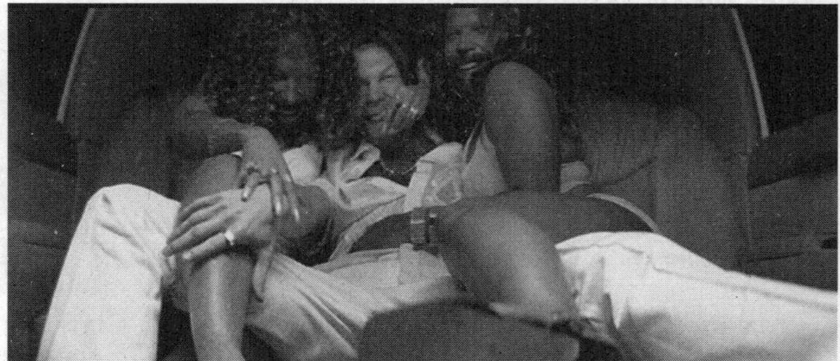

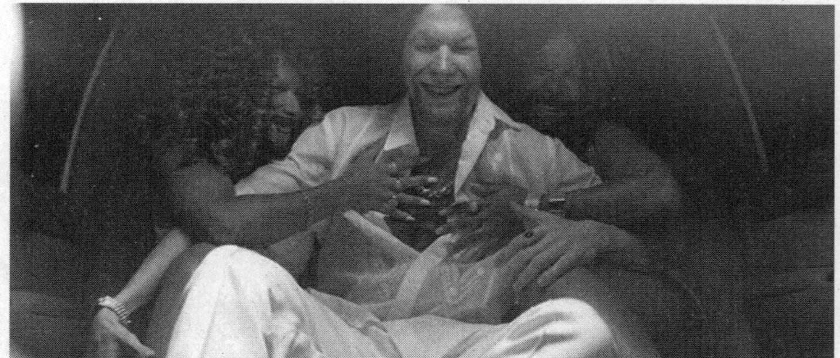

7.3   Chris Cunningham, *Windowlicker*, music video stills, 1999

nihilistic work both astonishes and repulses Gates, who sees in it a 'stew of life-denying imagery'.[2] At the time, I thought this was a perfect match. Casting myself as Gates, Cunningham became Dunkle: the doubting critic encountering the shock-tactic genius of music-video. Now I'm not so sure. After further viewing, it strikes me that there's more to Cunningham's body of work than first met my eye.

Bodies abound in Cunningham's films. Bodies human, android and unearthly. Bodies brutalized, metamorphosed and annihilated. In fact, Cunningham makes a point of transforming the artists he works with into his own creatures. Richard James (a.k.a. Aphex Twin), one of Cunningham's steadfast collaborators, has found himself reincarnated as, successively, a bunch of marauding kids (*Come to Daddy*), a high-rollin' pimp playa, and a bevy of ass-shaking showgirls (*Windowlicker*) all by means of a leering, slightly malevolent mask modelled on the musician's own features. In Cunningham's promo for *All is Full of Love*, Björk becomes a hormonal android, a desiring-machine engaged in a spot of production-line robot-love. Beth Gibbons is indeed present in Cunningham's promo for the Portishead song *Only You*, but she is held in watery limbo, her hair waving like fronds of weeds across her face, the viewer's attention displaced onto the sub-aquatic

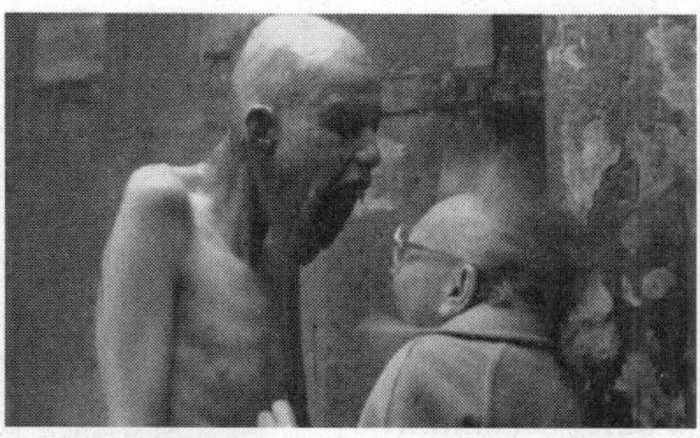

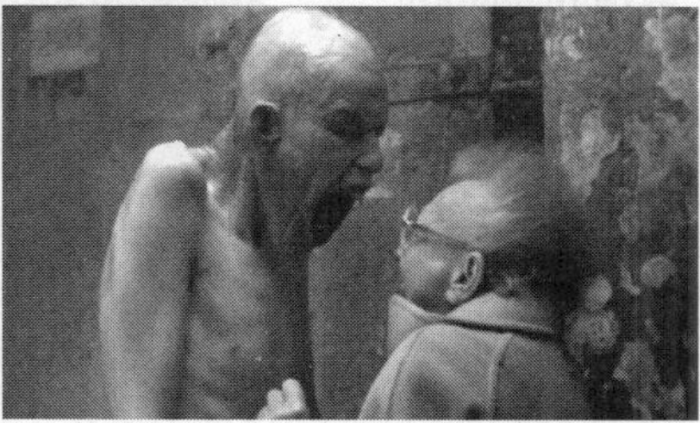

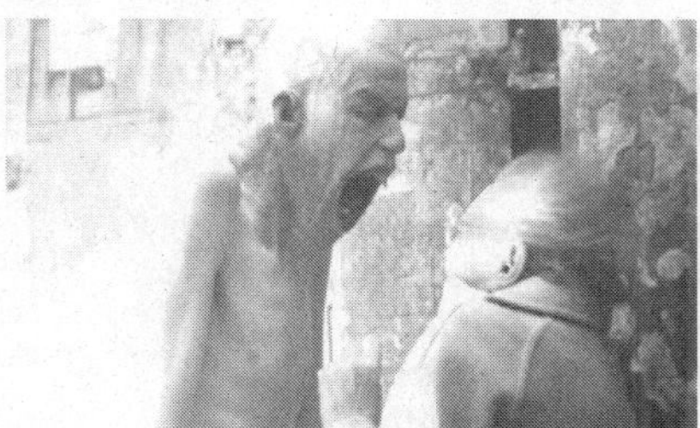

7.4   Chris Cunningham, *Come To Daddy*, music video stills, 1997

[2] Sarah Kent, *Venice Biennale Catalogue 2000-2001*, p. 262.

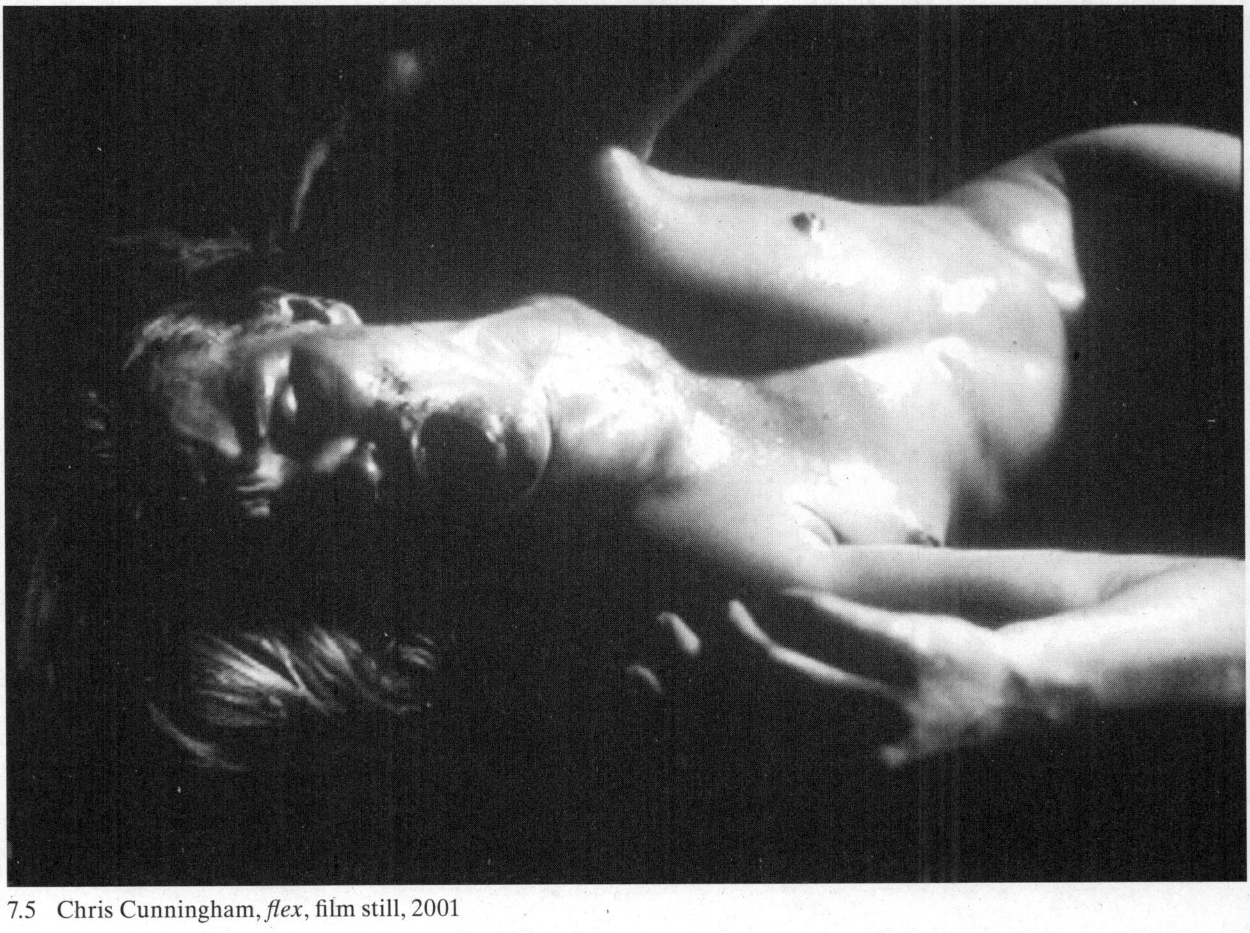

7.5   Chris Cunningham, *flex*, film still, 2001

acrobatics of a young boy, the object of her yearning attentions.

Cunningham not only transforms the body in the image but also the body of the image, that haptic combination of light, sound, movement and forms that Laura Marks has called 'the skin of the film'.[3] In Cunningham's hands, this 'skin' buckles, warps and yields. Music is a power, an elemental force that washes over, seeps into and dilates the body of the image. Perhaps it's not surprising that the work is pervaded by images of elemental powers, from a gravity-defying water-world (*Only You*) to an abstract void charged by a mysterious light (*flex*). Technology, too, has become elemental in Cunningham's world. Robotic fluids becomes bodily emissions. Television static is given demonic substance. But more than that, Cunningham's CGI interventions within the bodies themselves are marked in interesting ways. Their stylization shows us that technology is a power that intervenes in the flesh, in the brittle mineral body of a nomadic African going to pieces in Manhattan (Cunningham's promo for *Afrika (Shox)* by Leftfield), in cyborg-Björk's still-elfin humanity. But most of all in *Monkey Drummer* (2001), which looks like it escaped from the same scientific research institute as the other Cunningham creatures. A functioning assembly of fur, electronics and high-tensile steel, *Monkey*

[3] Laura U. Marks, *The Skin of the Film: Intercultural Cinema, Embodiment and the Senses*, Duke University Press, Durham and London, 2000.

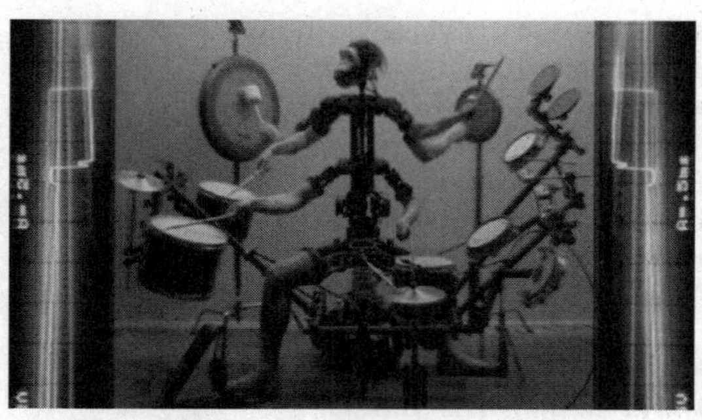

7.6 Chris Cunningham, *Monkey Drummer*, video still, 2001

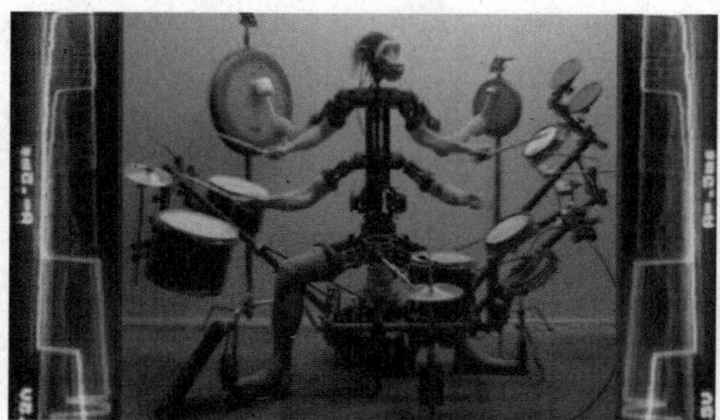

7.7 Chris Cunningham, *Monkey Drummer*, video still, 2001

Drummer is a metronome-accurate beat keeper running on the elemental power of rhythm, three pairs of industrial-robotic piston arms servicing impossible poly-rhythmic combinations. A pastiche-update of the sideshow amusement of days long gone, the piece is also a joke on techno music's propensity to severely alienate through its ultra-industrial beats-per-minute production.

In his music videos Cunningham treats the 'skin' of the film as a beat-driven surface through which the editing creates a perceptual shudder that, at times, is almost comically exact (for example, his promo for Squarepusher's *Come On, My Selecta*) and timed to the rhythmic micro-second. This adds another dimension to the director's evident practical and technical fascination with matters anatomical: it's the film itself that shifts and skitters with the hyper-kinetic rhythms created by many of the musicians Cunningham has chosen to work with. It's worth making the point that in his work with techno-artists such as Aphex Twin, Squarepusher and Autechre, Cunningham has created a visual repertoire of rhythmic effects using digital editing tools not far removed from the those that allow his chosen musicians to create their complex sound-worlds. In these terms, *Monkey Drummer* (first shown at the 2001 Venice Biennale) stands (or sits and plays) as a laconic declaration of artistic

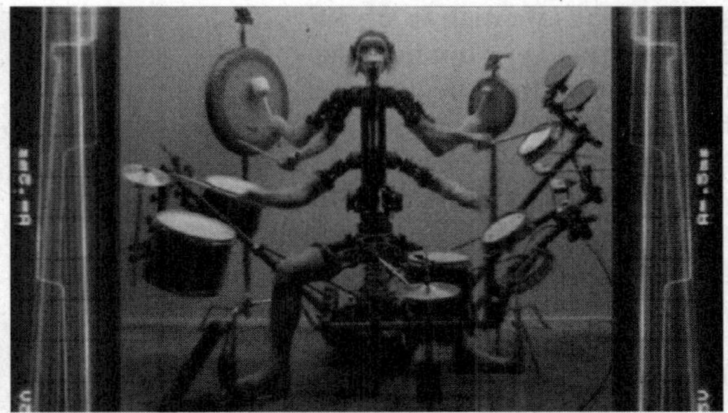

7.8 Chris Cunningham, *Monkey Drummer*, video still, 2001

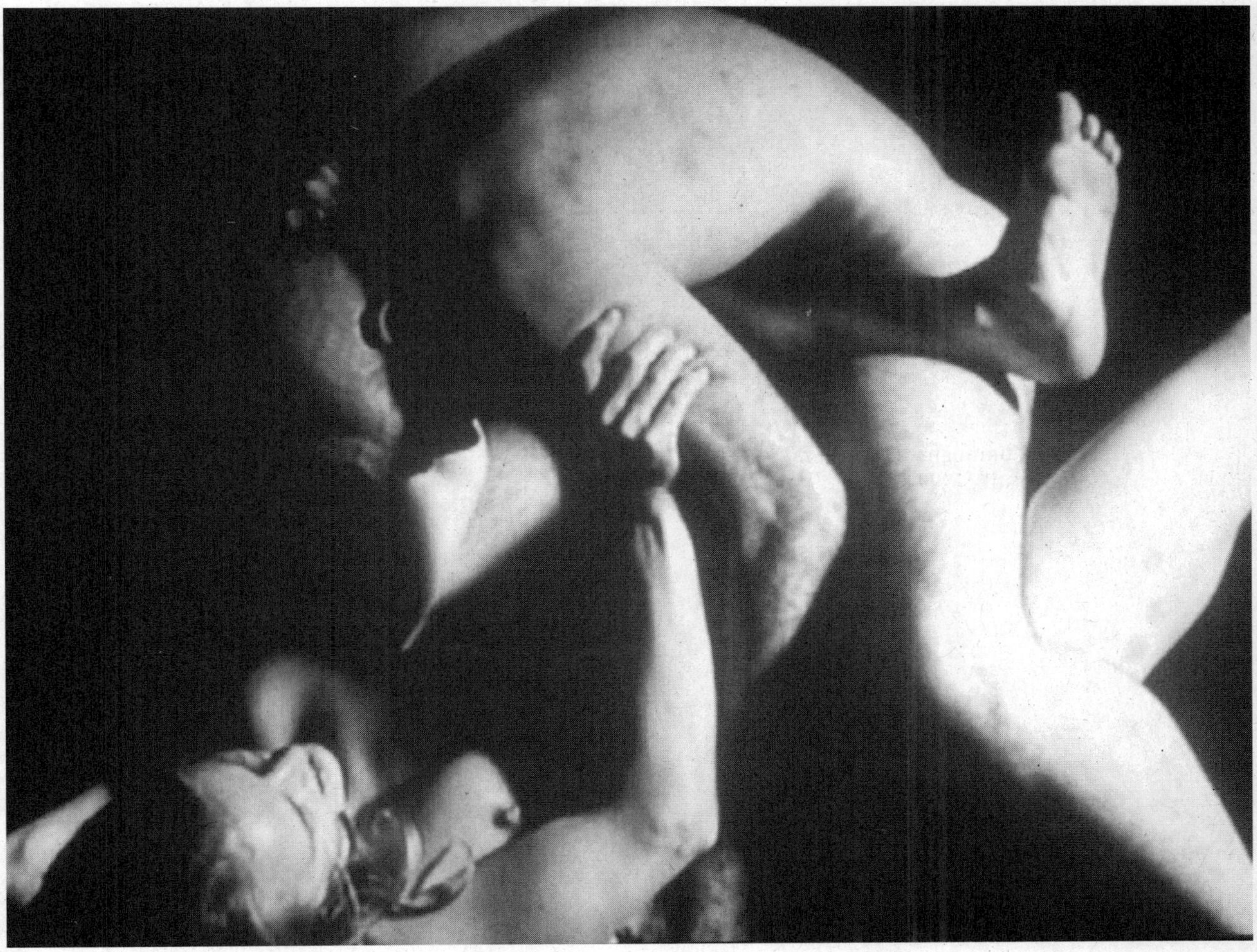

7.9   Chris Cunningham, *flex*, film still, 2001

intent: half-animal, half-machine, entirely driven by the rhythm.

Of course, this feature is most evident in Cunningham's work with techno and drum-and-bass artists, but is also present in a piece as relatively muted and tranquil as *All is Full of Love*. Cinematographer John Lynch, who shot the promo, recently described Cunningham's technical perfectionism when he explained in an interview how the director wanted to incorporate lens flare and the flicker of fluorescent light into the rhythmic texture of the piece: 'The fluorescent lights presented their own problems, because we couldn't dim them; we had about eight different people on the switches turning them on and off. Chris pays a great deal of attention to the photographic choreography, mapping out the visuals almost like music. In that regard we really had to work out the sequence and duration of the fluorescents' flicker throughout the song. We had to establish whether they would flicker all of the time, how long they'd stay on, where they'd flicker, when they went off, and so on.'[4]

Monkey Drummer and *flex* together represent a further development in the trajectory of Cunningham's career. Having started out as a model maker on such films as Clive Barker's *Nightbreed* (1990), David Fincher's *Alien 3*

[4] Christopher Probst, 'Amorous Androids', *American Cinematographer*, February 2000, p. 111.

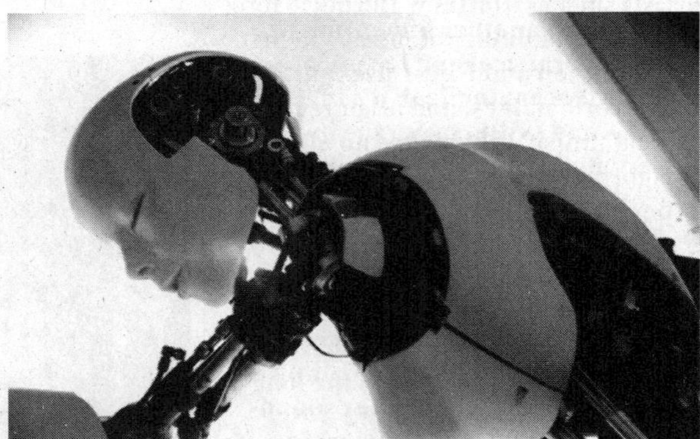

7.10 Chris Cunningham, *All is Full of Love*, music video still, 1999

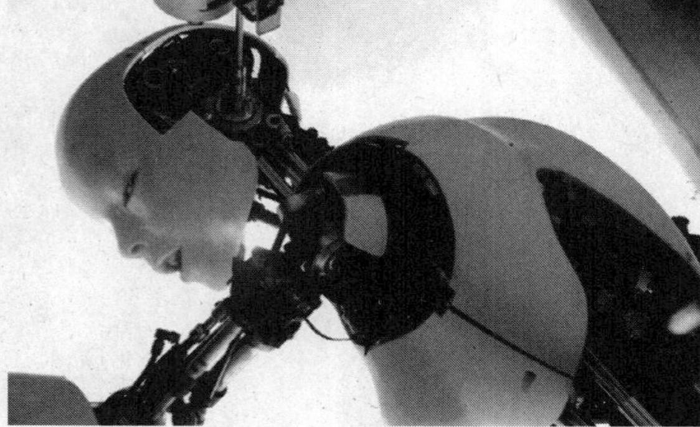

7.11 Chris Cunningham, *All is Full of Love*, music video still, 1999

(1992) and Danny Cannon's *Judge Dredd* (1995), as well as having laboured on designs for the robot child for Stanley Kubrick's aborted *AI* project, Cunningham then moved into music promos and commercials. With his most recent work, Cunningham has been courted by the art world. Is this development really as surprising as some commentators have suggested? In the coverage of *flex*, much was made of Cunningham's 'commercial' background, his evident ease and facility with expensive special effects, as though the moving image in the gallery still needed to be low-tech and low-res for the sake of 'authenticity'. Such misgivings appear misplaced. What about other artists such as Matthew Barney, Doug Aitken (himself a former maker of music videos), Mark Lewis, Isaac Julien and Eije-Liise Ahtilla? All are major art-world names, and all pursue a degree of technical 'finish' that makes no effort to conceal the rigorous professionalism of their work. What is perhaps most surprising about Cunningham's trajectory is that he has travelled from being a SFX 'backroom boy', via promo-video acclaim, into the art world without having made a feature film. Surprising, because his name has been attached for some time now to the long-awaited film version of William Gibson's novel *Neuromancer*. As with the other artist-filmmakers mentioned above, it is possible to speculate that the art world has become a de facto development

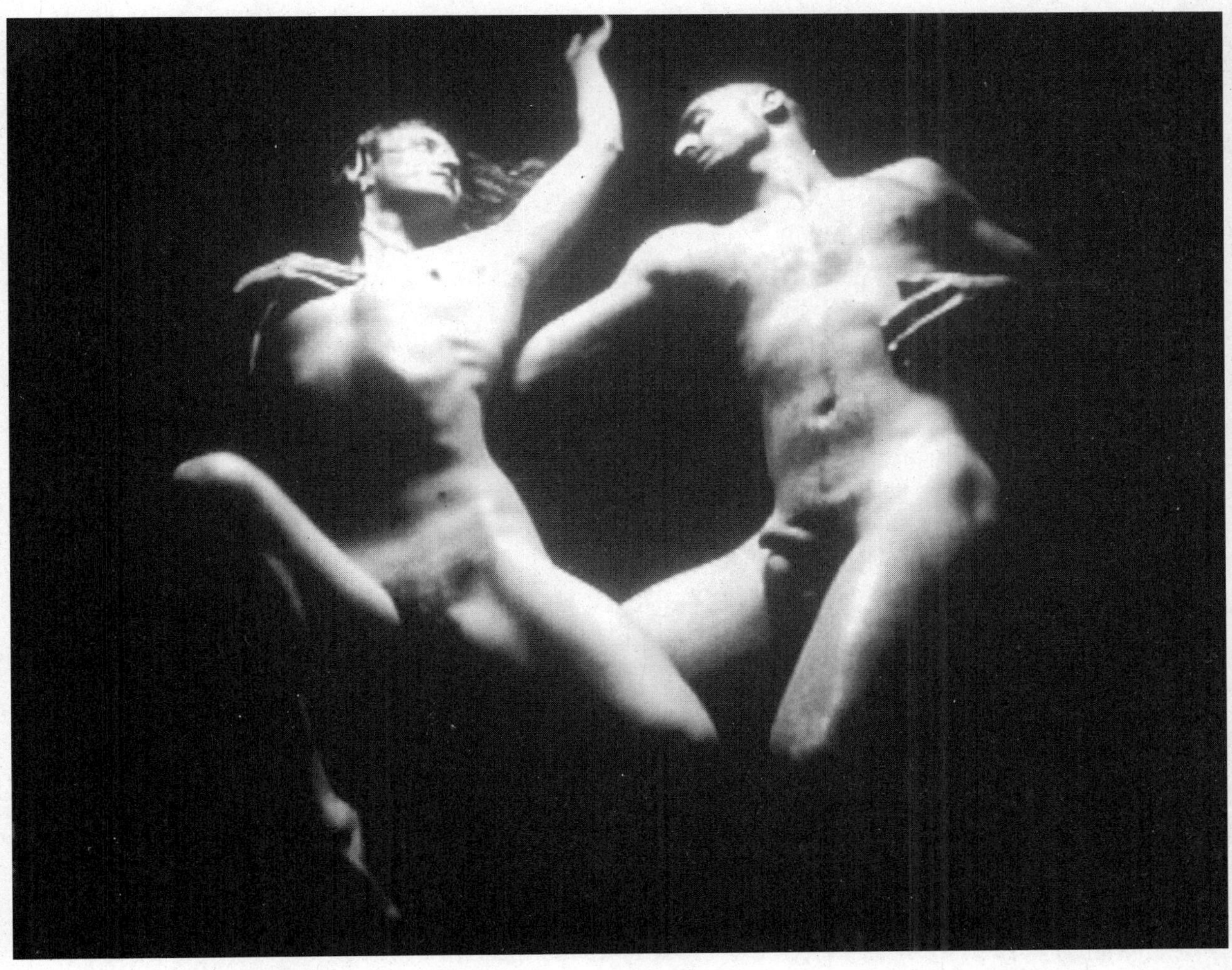

7.12 Chris Cunningham, *flex*, film still, 2001

space for filmmakers who want to work but whose national film industries are deeply conservative and unadventurous, certainly the case in the UK.

In *flex*, Cunningham made a 15-minute film that was screened in its own auditorium as part of the show *Apocalypse: Beauty and Horror in Contemporary Art* at the Royal Academy, London, in 2000. And he got to sculpt in the most interesting medium of all: time. The title condenses the elements of the piece – sex, fluid and flux – with the added suggestion of muscles rippling. A man and a woman wound together and held in Bill Viola-like amniotic suspension are struck by a mysterious light. They fight and fuck, hard and fast enough for *flex* to have carried age restrictions when it was shown at the Royal Academy. Alex Barber's camera renders the bodies as landscapes of mottled, ugly flesh as expressive as a Francis Bacon meat-puppet model, and Richard James scored the film with a sound-track that is equal parts *musique concrète* and percussive assault. While its production values are what one would expect of a high-end facilities house, it's a thoroughly otherworldly piece of work. Too much so, rumour has it, for its original commissioners Film Four, who passed it on to the Anthony d'Offay gallery. Further proof that the gallery circuit is increasingly home to work British film and television is too lily-livered to look at.

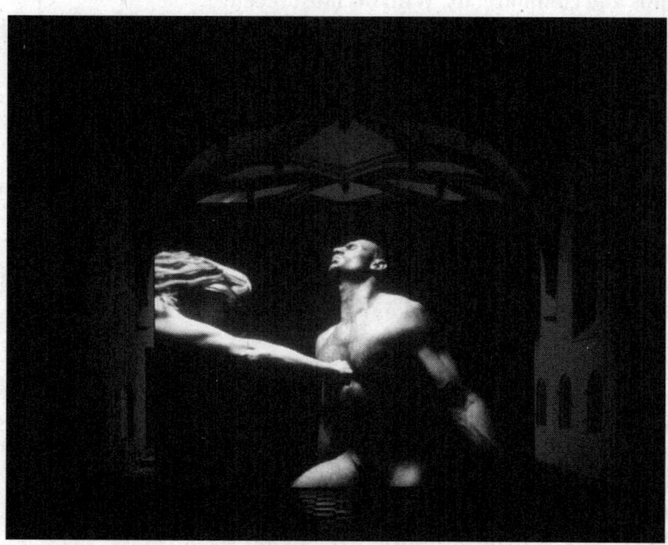

7.13 Chris Cunningham, *flex*, installation view, 2001

Its sensational aspects aside, the most interesting thing about *flex* is its serpentine sense of time: its 15 minutes are like a life cycle, a primal human wildlife film shot like an abstract horror movie.

Cunningham's work with 'the body' marks him out as an artist whose integration into the gallery circuit isn't overly problematic. But if the body is his subject, his register is that of 'the sublime'. Or, rather, a contemporary version of it. Drawing on Joseph Addison, Edmund Burke and, of course, Kant, the critic Scott Bukatman describes 'the sublime' as 'constituted through the combined sensations of astonishment, terror and awe that occur through the revelation of a power greater by far than the human'.[5] And in Cunningham's work this power 'greater by far than the human' is technological. Bukatman also observes that the ancient and highly useful concept of 'the sublime' now overlaps as readily with the world of technology as it has ever done with the world of nature: 'The sublime not only points back towards an historical past; it also holds out the promise of self-fulfilment and technological transcendence in an imaginable near future. Under the terms of the sublime, technology is divorced from its sociological, rationalist underpinnings to become a technology without technocracy, a technology beyond the scope of human control. There is thus an inevitability to the fact of technological

[5] Scott Bukatman, 'The Artificial Infinite: On Special Effects and The Sublime', in Annette Kuhn (ed.), *Alien Zone 2: The Spaces of Science Fiction Cinema*, Verso, London and New York, 1999, p. 255.

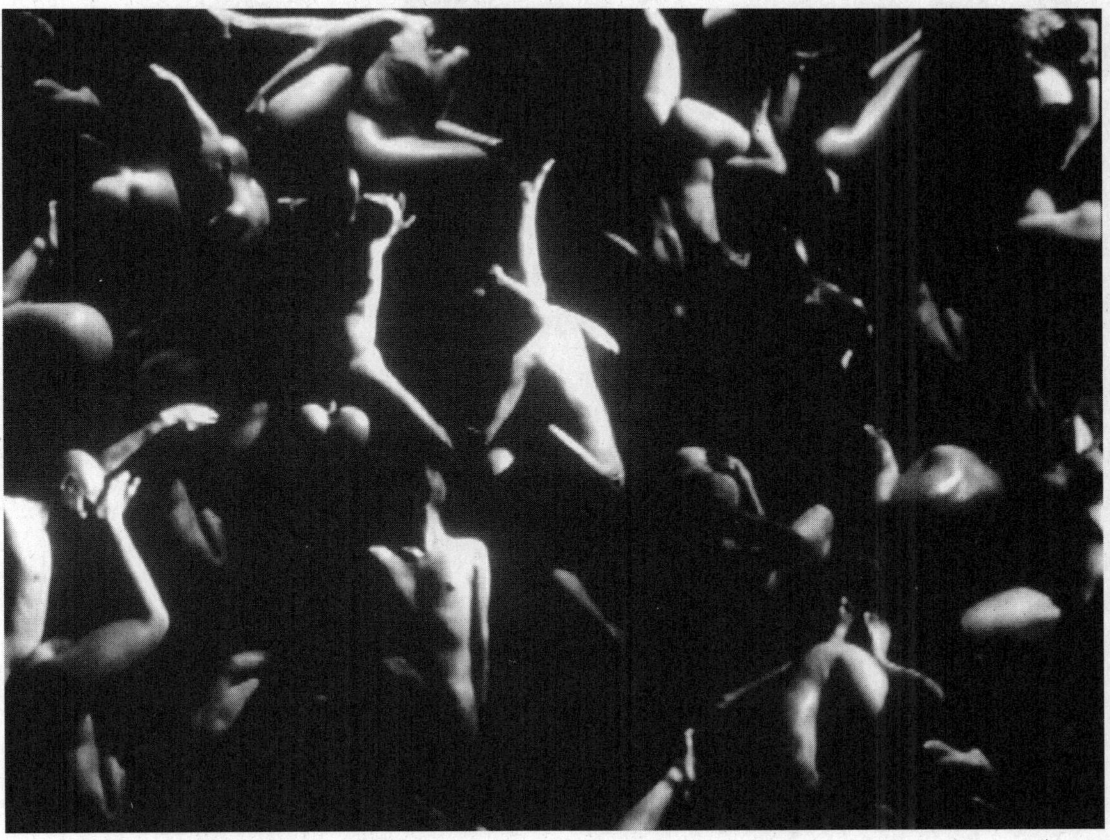

7.14 Chris Cunningham, *flex*, film still, 2001

progress, and thus accommodation becomes the one valid response.'[6] **Remodelled by Cunningham for the twenty-first century, 'the sublime' – hitherto a matter, following Goethe, of 'beauty and terror'**[7]**– becomes the interplay of body and technology. And, judging from Cunningham's work, the ongoing 'accommodation' of one by the other is not without its price.**

Yet for all the technical perfection, all the undoubted polish and finesse Cunningham's work displays, there remains something ultimately adolescent in its vision. 'Adolescent' in its attraction to the extremities of sex and violence, in its attitudes towards men and women (from *flex* to *Windowlicker* to *Come to Daddy*), and it is not yet developed enough to be interestingly perverse, nor yet mature enough a vision to betray that crucial quality – a thought, an ethic, a world view (but maybe it's a bit much to expect Cunningham to be Jean Renoir!) – beyond the easy 'subversiveness' much prized by TV programmers and, increasingly, curators. But, who knows, maybe if (or when) Cunningham finally gets to grips with long-form narrative, his much-vaunted 'style' will finally give some heart to all those sublime bodies?

[6] Ibid, p. 270.
[7] Rainer Maria Rilke's famous lines from the *Duino Elegies* serve as a poetic definition of 'the sublime': 'Beauty is only/ the first touch of terror/ we can still bear/ and it awes us so much/ because it so coolly/ disdains to destroy us.'

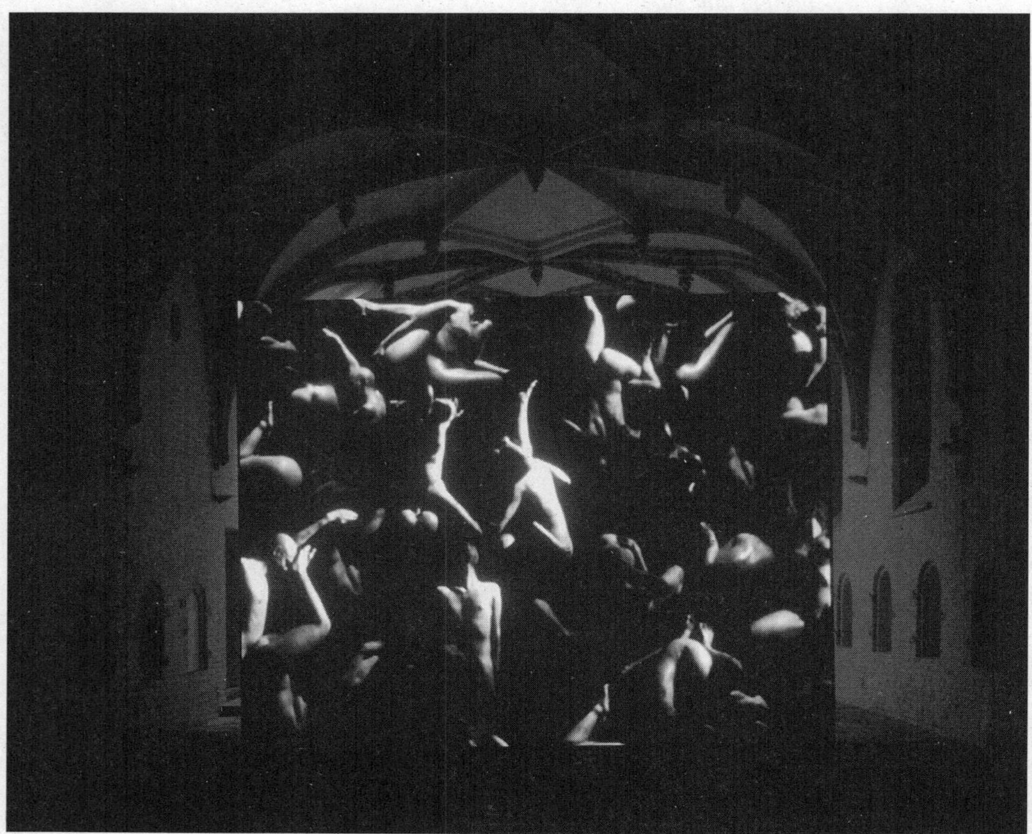

7.15 Chris Cunningham, *flex*, installation view, 2001

Edwin Carels: Unguarded Moments

They're off once more. The school gates have reopened, streets and squares are empty and everyday life has regained its usual, familiar rhythm. Children and adolescents are divided in groups, put in classrooms and served mouthfuls of knowledge. Family life, too, automatically becomes somewhat more orderly. But how does the child feel, in what activities does he or she engage after school and before supper? Playing, enjoying sweets, reading comics, watching television, indulging in narcissism, listening to music: these favourite pastimes allow children to enjoy unguarded moments of pleasure. Just for a brief moment the ever-watchful eyes within the surroundings of the school and the conditioning control of parents at home are absent. And therefore inhibitions as well. The regulating social perspective is exchanged for unashamed spontaneity. Or so it seems. For even during their free time, young people are constantly steered by trend watchers, pop idols and all sorts of trend setters. Youth is undoubtedly the most coveted and least respected stage in human existence. It is not easy to be young if you are constantly being watched, focused upon, charted. To what extent are their growing self-awareness and self-image defined by children themselves?

I want your soul,
I will eat your soul.
I want your soul,
I will eat your soul.
I want your soul,
I will eat your soul.
I want your soul,
I will eat your soul.
Come to daddy.
Come to daddy.
Come to daddy.
Come to daddy.
Come to daddy.
Come to daddy.
Come to daddy.
I want your soul,
I want your soul,
I want your soul,
Aargh.

(Aphex Twin)

Breeding Grounds?

The Flemish expression *terug van school* ('back from school') sounds somewhat odd in the Netherlands. The Dutch prefer the expression *school is uit* ('school's out'); this, however, has a somewhat different connotation. 'Out' implies an element of escape. It even sounds rather jubilant, as if an ordeal has come to an end. Yet school is never 'over and out'. School holidays and free time only exist as the negation of school time. Salvation is always relative. 'Back from school', on the contrary, seems to suggest that some sort of temporarily autonomous zone comes into being during the action of movement, rather than inside or outside school. Walking, cycling or taking the bus, the child finds itself between two destinations, between two strictly separated zones ruled by definite expectations. Compared to these, traffic rules seems trivial. Straying from the direct route, making a detour to pop in at the candy shop, the toy shop, the youth club, to sit on a bench in the park: it's all too easy, since the parents are not home yet. Unguarded moments. All children are in search of them, whether there are cameras in the street or not.

Though 'back from school' is less about behaviour at school than about the stubborn self-determination or absent-mindedness of

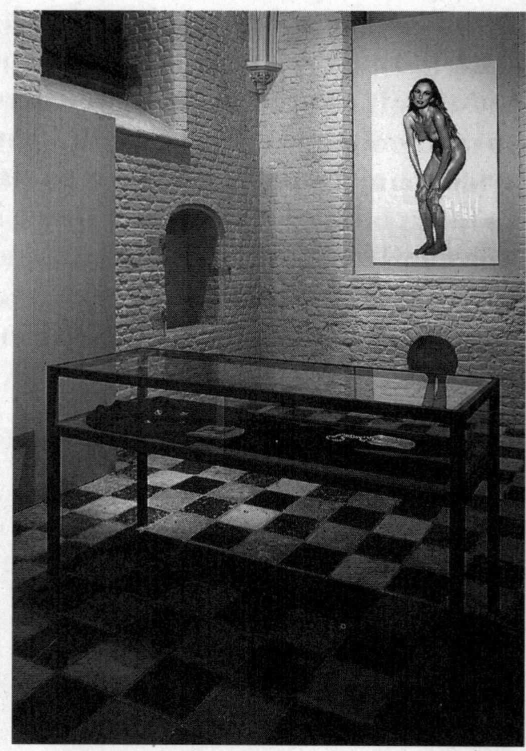

8.1 *Terug van School*, installation view, 2000

children during leisure time, it is obvious that the surroundings in which children grow up play an important role. Something similar applies to the elaboration of the concept of the exhibition *Back from School*. The project was inspired not just by current events but by its location as well. Middelburg is situated near the border with Catholic Belgium, yet its character is very different, because Dutch Reformed culture is very present in village life. A spokesperson from the National School Museum in Rotterdam once pointed out that even today, Protestant teachers on guided tours of the museum frequently ask to skip the first 'stops', because the images are too explicit (i.e. explicitly Catholic). The set-up of *School in Those Days* in Ghent, meanwhile, is conspicuously more chaotic, more Baroque, more lively, too, than the rigid reconstruction in Rotterdam. Though European commissions work hard to bring some uniformity in (higher) education and national ministries impose a standard curriculum on schools, small and even large differences remain in methodology and organization. Much as Claes Oldenburg's *Soft Alphabet* invites the viewer to touch instead of read, in the attics of school museums we find often strange instruments and teaching aids that bear witness to the most peculiar teaching methods of the past. Each child bears not only the mark of the educational system, but also of

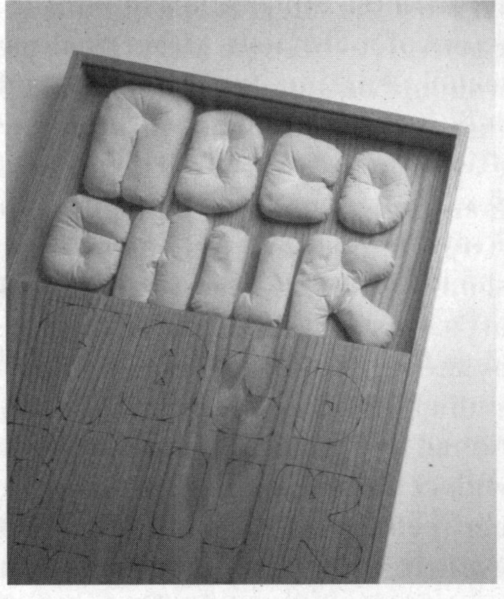

8.2   Claes Oldenburg, *Soft Alphabet*, 1978

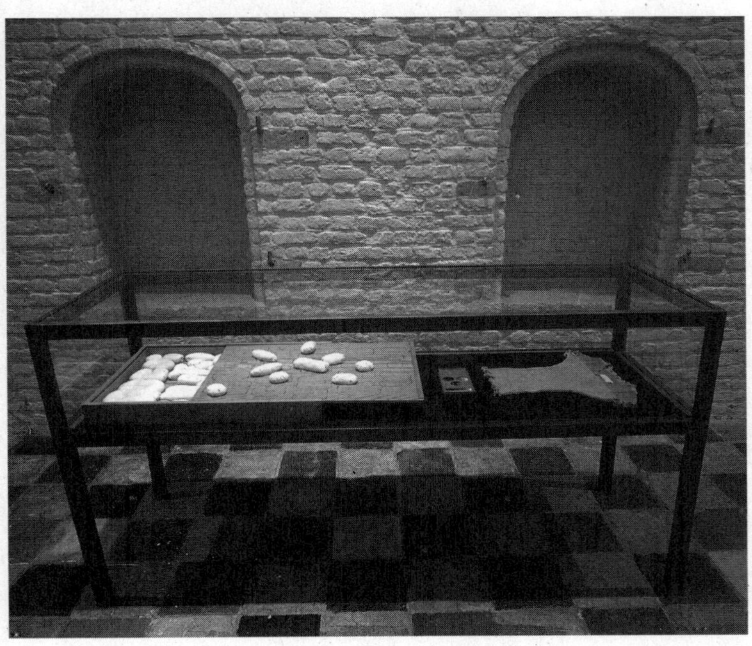

8.3   *Terug van School*, installation view, 2000

the particular school he or she went to and of his or
her individual teachers. Single-sex boarding schools
are almost unheard of in the Netherlands, while in
some regions of Flanders this 'unworldly system'
still exerts a direct influence.

A Ban on Public Gatherings?

Just as the slightly deviating grammar of its title
elicits both recognition and alienation, the predom-
inance of Flemish artists in the exhibition may
have peculiar connotations. They will, of course,
not bring about a culture shock in Dutch visitors.
At most the effect is one of mild alienation, which
causes us to become alert. To dispel any doubt,
it should be noted that *Back from School* is a
thematic exhibition of works that were created
earlier, in a different context. The exhibition is
therefore a somewhat artificial collection of objects.
Any reference to a Belgian or Flemish School
should therefore be avoided, though individual
artists are, of course, influenced to a certain extent
by the spirit of the times and often bear witness to
similar preoccupations. The public, in turn, conditioned by a particular logic, approaches the artists
with certain expectations, making certain links
between the artists. These, too, had better be made
explicit. Danny Devos is quite illustrative in this
context. Because of the Dutroux affair his *Perpe-*

trators of Killings suddenly attracted significant interest from the public and the media. Until then, the fact that this artist had been preoccupied for quite a number of years with the theme of serial killers as a metaphor for the artist had largely gone unnoticed. In Koen Theys's series of photographs *Kindergarten*, inevitably, discreetly, the faces of the missing children from the Dutroux era – once the focus of all media – appear. The artists visited dozens of nursery schools in and around Brussels, asking to photograph children's classroom creations. What inspired the artist was not the political climate prevailing at the time, but rather the observation that certain cultural codes are imposed on very young children. He was intrigued, not by the big bad wolf, but by the eerie seriality with which children are asked to produce sheep by gluing cotton clouds on identical pre-drawn shapes. Once Theys even heard a teacher address the children with these (literal) words: 'Now, children, be quiet and get on with your drawing. There's an artist in the classroom.'

Childish?

For the exhibition *Back from School* we have selected artists who work with the camera (photographs, video, film) or who use contemporary audiovisual media (CD-ROM, photocopies).

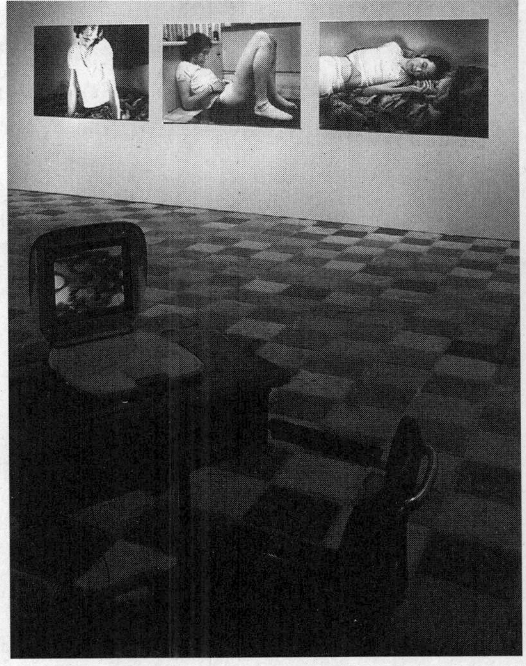

8.4   *Terug van School*, installation view, 2000

Paintings are made using a paint box on a video screen; the only sculpture is Duane Hanson's *Seated Child*, portraying a life-size, hyper-realistic child. Even a simple raincoat, styled like a grey cloak (a work by Martin Margiela) is a simulation: an almost photographic reproduction of a model that was sold a hundred years ago. Inevitably, contemporary artists have to engage in a dialogue with the media and all manifestations of the visual culture in which they are immersed in order to measure their own images. That is also true for our children. They, too, are constantly interacting with culture, either rejecting it or identifying with it. The metaphor of the blank sheet of paper – to which we referred in days gone by – had better be replaced by that of an overexposed sheet of photo paper that leaves us in doubt as to what image will finally emerge.

It is an ever-more vicious circle as the child turns to the media to construct his or her self-image, while the media relies more heavily upon canonical images of childlike innocence. In Belgium in recent years the idea of children's innocence has been cultivated to an especially high degree while equally strong fantasies have been created about how this innocence might be lost. What was going through the heads of those hundreds of thousands of people who took part in the White March on 20

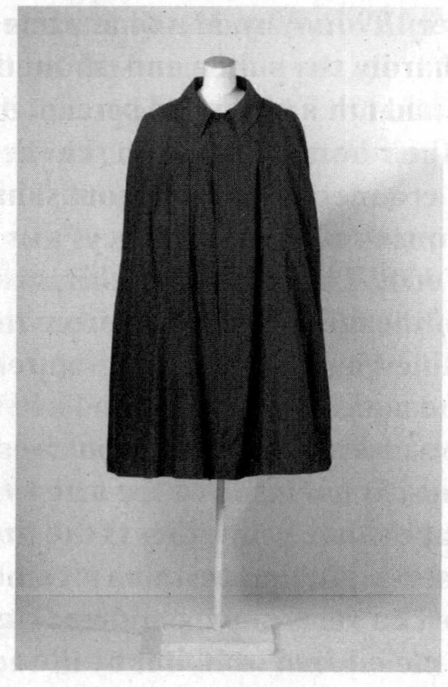

8.5   Maison Martin Margiela, Remake of a cavalry cape, Fall/Winter collection 1995–1996

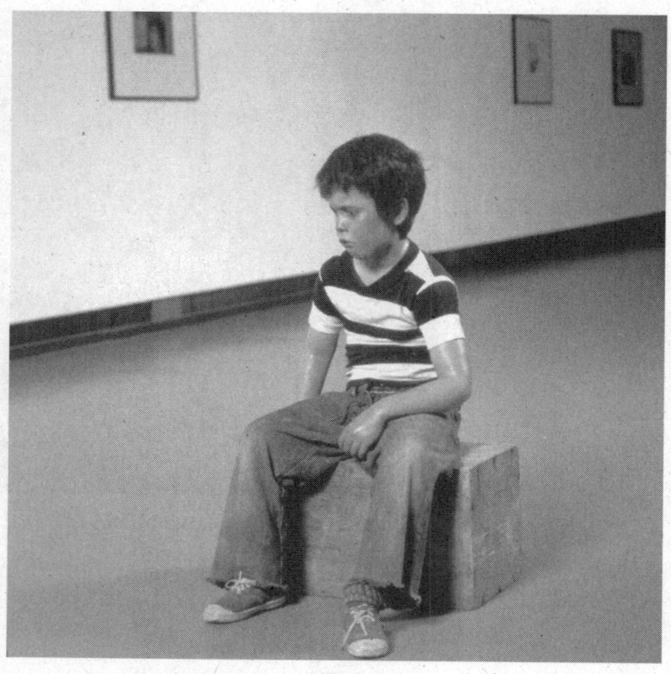

8.6   Duane Hanson, *Seated Child*, 1974

October 1996? Why, at a time when the police had made hardly any statements about their inquiry and gossip had not had time to spread, did they drag along their own unsuspecting children on crammed trains, crowded streets, without sanitary facilities, not knowing when the agony of this pilgrimage would end? The result of this mass hysteria certainly was that a whole generation of children and adolescents became prematurely acquainted with the notion of sex crimes. The White March (a march without slogans, without message, without content) did not bring about a new October Revolution, but it did offer the distressed public a gigantic screen upon which it could project all sorts of fantasies of personal anxiety, focusing on the so-called face of evil, the big bad wolf: Marc Dutroux. The safety issue rose to the top of the agenda as an item for both right-wing and left-wing parties to use to court voters. As if more uniformed police in the streets would lead to fewer sexual assaults in private settings.

Model Child?

Often, parents themselves impose an accelerated hyper-adulthood on their children, forcing responsibilities upon them and demanding that they behave like moral human beings. They treat their children as proto-adults, bring them in situations that are alien to their nature as children. Then they are flabbergasted because something 'bad' (some grown-up evil) has happened to their so-called innocent children. A notorious example in this respect is the case of JonBenét Ramsey. At the age of six, the girl had already participated in countless beauty contests when she was found murdered in the basement of her home on Christmas Day 1996. The case remains unsolved up to this very day. Because of unlimited access to television, children learn to understand things at an earlier age; they absorb images and words, learn to call things by their name. The blank sheet to which we referred earlier does not exist, of course. Every child discovers physical impulses at an early age, impulses adults are all too ready to catalogue and interpret. For his installation *Once And For All*, Patrick Van Caeckenbergh used a huge number of pornographic magazines, putting an overabundance of uncovered, undefined shades of skin on view. The serene, serial presentation of these naked fragments of truth – which are as individual as they are anonymous – results in a both cheerful and depressing 'environment' that is at once a paradise and a prison.

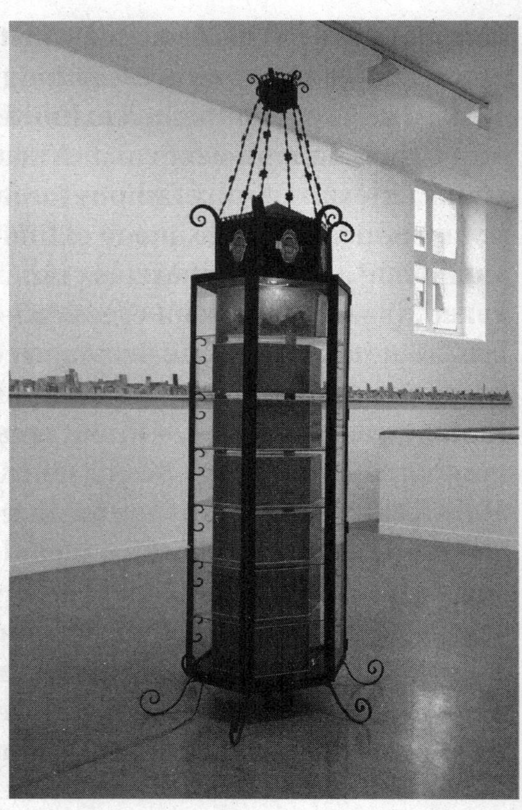

8.7   Patrick Van Caeckenbergh, *Collection de peaux*, 1991–1992

The Dutroux affair was news all around the world. Never before has paedophilia seemed to dominate the collective consciousness to such

an extent. This is confirmed, for instance, by the
fuss a single – not even eye-catching – photograph
created in 1999 at a themed exhibition organized as
part of the Holland Festival in Amsterdam. Even
in this city, supposedly famous for its tolerance,
it was suddenly impossible to exhibit a picture
portraying a father holding his son in his arms, both
naked, the father with an erection – a picture from
1962 that had been published many times before.
For that matter, the urge to see a pornographic
content was certainly not in the mind of its maker
(photographer Walter Chapell), but in the minds
of the camera crew of a scandal-mongering local
news station. Rarely are the media (producers
and consumers) aware of the perversity of the
perspective from which they present the 'youth
issue'. 'Loving you is easy 'cause you're beautiful,'
we hear the artist Luc Compernol singing on his
video *Covering #1: Beautiful*, emphatically out of
tune as we see the image of two flowers in a vase
being drawn over with crude, childish force. At
the CAPC in Bordeaux, France, the organizers
of the exhibition *Presumed Innocent* had to make
do without posters in the streets, without info for
the public on the official website; furthermore, a
number of photographs had to be removed from
the catalogue. As civil servants, the curators had to
watch over their exhibition about youth in contemporary art with constant vigilance. Three months

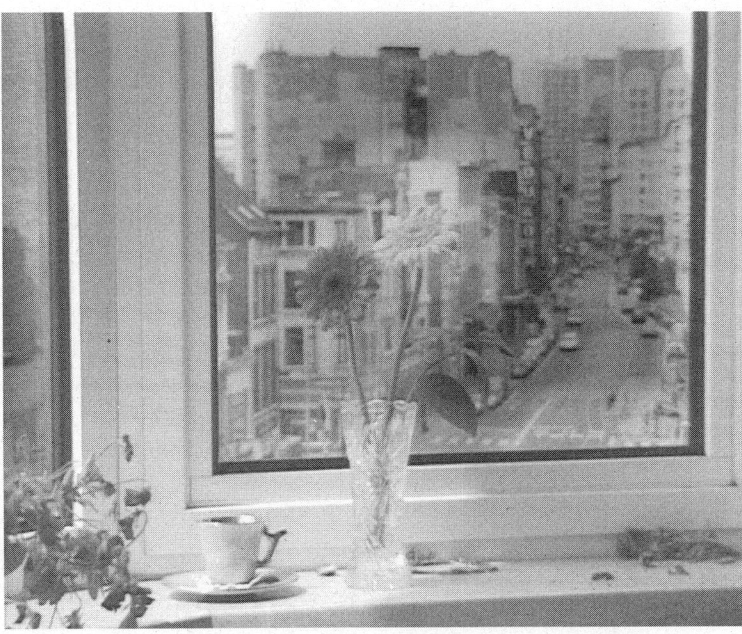

8.8   Luc Compernol, *Covering #1: Beautiful*, 1999

earlier, the so-called progressive quality newspaper *Libération* had published an exhaustive dossier on the subject of paedophilia that had greatly alarmed the city fathers. The city, popular with tourists, could not be cast in a bad light...

Sorrow and Misery?

It seems no summer can pass without paedophilia hitting the headlines. As more and more information channels operate on a profit basis, journalists are required to score continuously. No more off-season. The recipe for success is simple and primitive: to sell news, the public's curiosity has to be aroused, and nothing arouses more curiosity than a flirt with taboos. As paedophilia and incest are just about the only taboos that have not yet been completely exhausted, the media are constantly on paedo-watch, even hunting down alleged paedophiles. The difference between 'Name and Shame' campaigns by tabloids such as *The News of The World* and the coverage of the subject in so-called quality papers is merely superficial and actually irrelevant. No matter the approach, the fact is paedophilia is considered a headline item, and imagination is given free play.

Whereas in neighbouring countries these morbid affairs are considered criminal ephemera

and generally sink into oblivion quickly, in Belgium
the ghost of Dutroux keeps haunting the collective
consciousness. Why? Months after Dutroux's
arrest, against their better judgement (and for want
of better footage), television stations broadcast
the same scene countless times, one that was
burned like a diabolic mantra on our retinas: the
handcuffed child rapist is led down the steps of the
court and pushed in a police van, all in a matter
of seconds. *Dutroux descendant un escalier*, the
descent down the Neufchâteau staircase. Will this
image prove as enduring as Eisenstein's famous
sequence from *Battleship Potemkin*? Just as the
avant-garde revolutionary film maker manipulated
and magnified the passing of real time in his famous
montage of the Odessa steps in order to attain a
certain psychological effect, the editors of television
newscasts very efficiently turned an anecdotal
figure into an indelible archetype. This was
unadulterated experimental television: hundreds,
thousands of times in the course of a few weeks, the
same few seconds were broadcast. To ward off evil?
To fuel fear? Maybe for both reasons.

The surrealistically repeated staircase scene
is the only image that confronts us in a visually
interesting way with images of Dutroux in motion.
Apart from this scene, innumerable variations of
the 'face of evil' (with or without a beard) abound

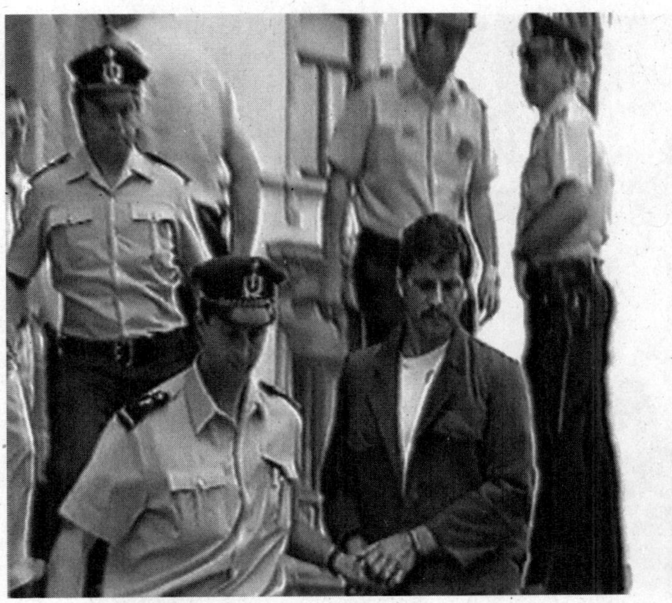

8.9   Television footage of Marc Dutroux's decent from the
Neufchâteau steps, 1998

in the printed media. Not until November 1999 did
the famous picture of Dutroux as a child emerge,
the picture reproduced on the invitation card and
the floor plan of the *Back from School* exhibition:
Marc Dutroux in Africa in the 1950s. The photograph was published in the cultural supplement to
De Financieel Economische Tijd (the only Belgian
quality newspaper that need not resort to sensationalism to keep its market share) on the occasion of a
review of the novels *Zwarte Tranen* ('Black Tears')
by Tom Lanoye and *Open gelijk een mond* ('Open
Like a Mouth') by Jeroen Olyslaeghers. Both are
ironic *romans à clef*, and it takes no great effort
– *verbum sap* – to recognize in them a caricature
of the events that followed the Dutroux scandal.
It is the sort of literature that parodies the media
but actually offers nothing beyond that. The two
novels do not dig particularly deep into the causes
of either the collective paranoia or the individual
pathology. The photograph gave rise to a more
interesting debate than the novels. The reason for
it not turning up sooner – though it could retrieved
freely from the Photo News image database – is
actually quite simple. The emotive image of the boy
who somewhat ruefully, aggrievedly, even rather
neglectedly, plays with a toppled wheelbarrow
confronted the press with an awkward paradox.
How could the monster Dutroux be associated with
this endearing child? An image like this arouses

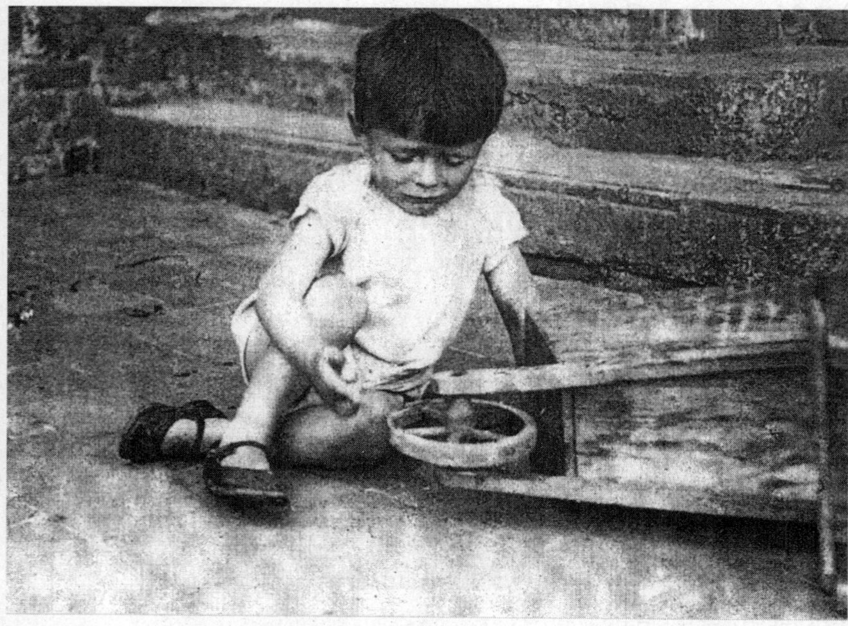

8.10 Marc Dutroux in Africa in the 1950's

an inconvenient sympathy that is hard to reconcile
with the simplistic, binary opposition between the
concept of the universal innocence of the child and
the unique, particular character of the paedophile
murderer. This might be experienced by the public
as a perceptual short-circuit. Is this the face of a
child rapist?

Sanctuary?

Would it still be possible to exhibit Marijke van
Warmerdam's *Handstand* (1992) in de United
States, now that the moral majority finds the
slightest pretext to wage war in even liberal New
York? Indeed, the cheerful girl, dressed in a white
skirt and knickers, doing a playful, endlessly
looped handstand against a blank wall, is just
the sort of thing for paedophilic voyeurs. Yet just
as 'beauty is in the eyes of the beholder', so is
the reverse. Innocent or the focus of a thousand
eyes: art is never viewed without preconceived
notions. Whoever looks for it can find a dubious
hidden meaning in the most innocent of works.
Furthermore, if even artists have to be subservient
to blunt ethical standards, there is really no more
need for art. It is precisely ambiguity and being-
somewhere-in-between that make a work of art
worth contemplating. Those who turn the sanctuary
of art into a battlefield are guilty of self-fulfilling

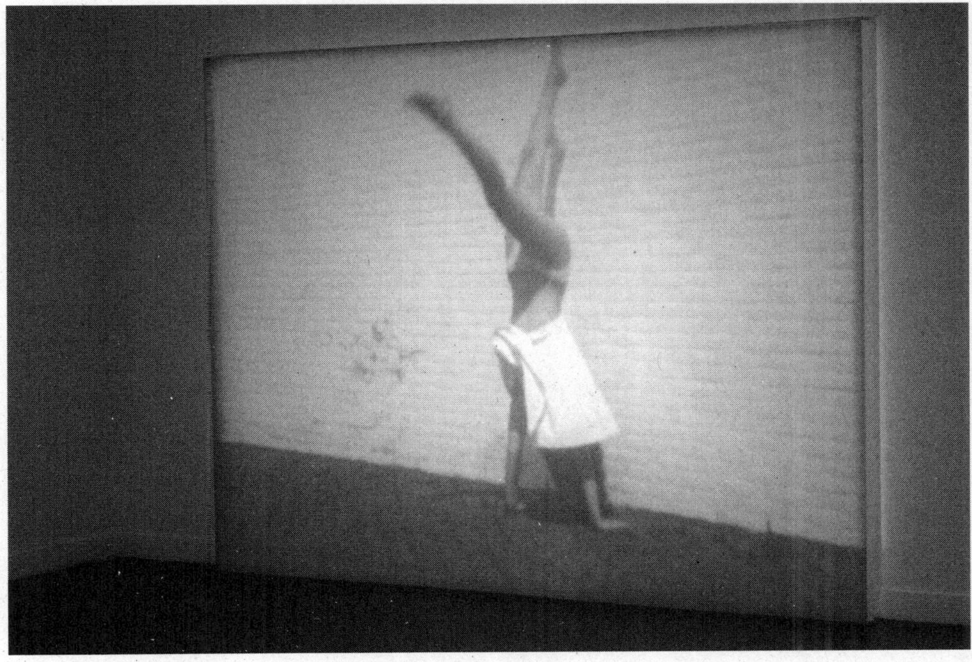

8.11 Marijke van Warmerdam, *Handstand*, 1992

prophecies about good and evil. The curators
of *Presumed Innocent* were therefore criticized
merely because of the title of their exhibition: they
had deliberately left out who is presumed innocent:
the child, the artist or the viewer? In his essay
'Presume Nothing: Thoughts on The (Endless) End
of Innocence', Joshua Decter expands on this idea:
'Whether intended from a legal, religious or moral
point of view, the concept of "innocence" presup-
poses the endorsement of a whole set of ethical and
social principles that are defined by a binary logic.
Either one is guilty, or one is not. Is it possible to be
a little innocent? Or a little guilty? The cultivation
of contrasts, or worse, the denial of part of reality,
is a tempting, convenient solution, in both the realm
of politics and education.'

While children are taught as much positivist
knowledge as possible, what they really are
curious about is their own physicality and that of
others. The chaos caused by bodies awakening,
by rampant emotions and social variables rarely
results in a predictably, homogeneously composed
classroom nowadays. According to a study which
surveyed 14,000 British children, one girl in six
reaches puberty by the age of eight. Educational
paradigms may be necessary, but the simple
division of nursery school, primary school and
secondary school gives us little to go on to. The

school system that is common to this day thrives on privileges and failures. It is a perfectly orchestrated theatre, with boxes that divide children along lines of social status, gender and ambition. At the same time it remains an arena that stimulates primitive passions, clan spirit, revolt and subjection. The function of the school is to order, to classify and to pass on knowledge. Discipline and punishment are essential. The titles and degrees schools hand out are a reliable measure to gauge future success and wealth. Too much trouble with hormones at secondary school may therefore have far-reaching consequences. Just as the looped *Handstand* confronts us with both playfulness and rigid order, *Back from School* confronts us with the conflict between rational and emotional development through the emphasis on the systematic, serial ordering (in the presentation of angular series) on the one hand, and on playful mobility (balls abounding everywhere) on the other hand. Robert Gober's red wax shoe figures both helplessly and proudly among some fashionable pieces of musical equipment. Who or what cast a spell on this little shoe? The unbridled zest for life of the child or the pre-programmed pseudo-energy of the pop industry? The truth, of course, is somewhere in the middle. Sports and games still remain the most pleasant method to mould our attention and energy, to achieve a certain attitude. Yet, some children

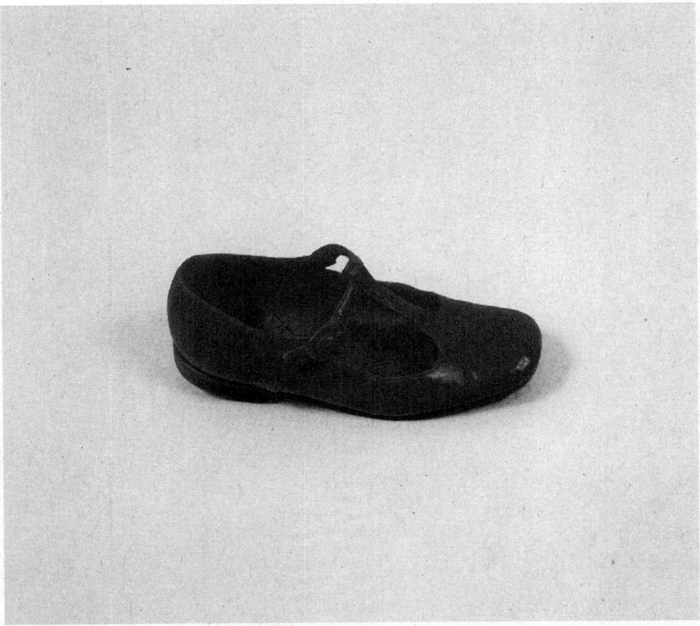

8.12 Robert Gober, *Untitled Shoe*, 1990

seem to experience the many gym apparatuses along their path as instruments of torture.

Learn Young?

The American artist Sharon Lockhart had some children on a staircase inside a school building re-enact a scene from a film by Truffaut. In the scene, a couple is about to kiss. She called the series of five photographs she distilled from this re-enactment *Audition Series*. Five times she has portrayed the same scene, each time with different actors. Probably none of these children exactly understood the artist's intention, let alone had ever heard of François Truffaut. Adults constantly project all sorts of images onto schoolchildren, often inspired by the noblest of intentions, but without the children having asked for this or being involved directly. And whereas children are allowed unrestricted access to the visual media, visual education remains the least urgent of the school's concerns – if the subject is taught at all.

The prevailing view of young children is usually a construct thought up by adults. Adolescence as a category has existed for scarcely two centuries. Some believe the concept to have been promoted by developments in photography, criminology, the medical profession and psychoanalysis.

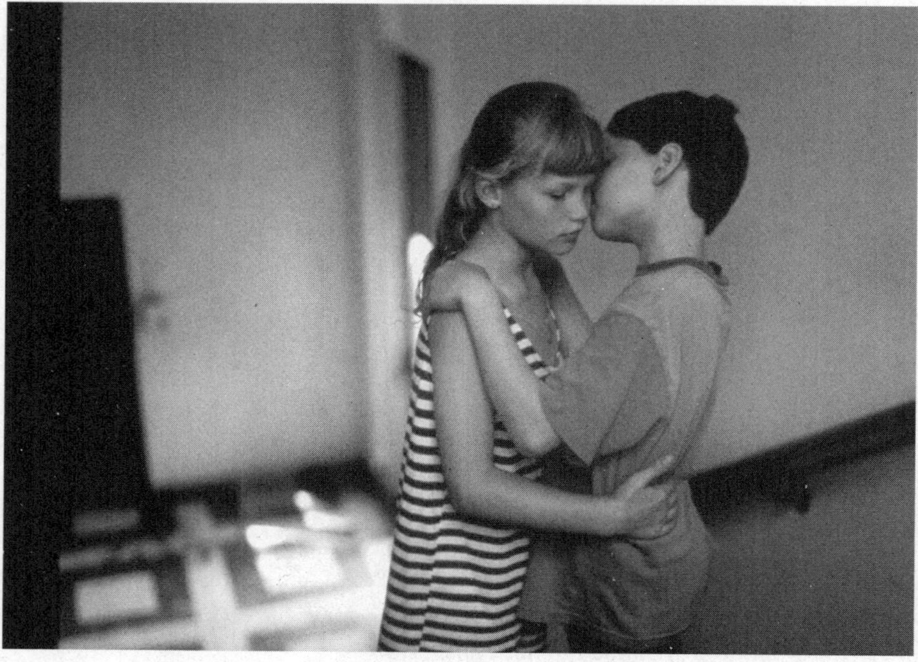

8.13 Sharon Lockhart, *Audition Two: Darija and Daniel*, 1994

Between 1780 and 1840 countless books on puberty and its accompanying physiological changes were published. In 1896 a chair in *hébéologie* (the science of puberty) was even established at the University of Paris. Adolescence is a Western invention that uses a conspicuous amount of terminology borrowed from the science of pathology, which might explain the connotation of puberty with 'crisis'. Until the arrival of psychoanalysis, adolescents were simply considered undisciplined and potential delinquents. Economic circumstances, too, favoured the promotion of puberty as a new category. The invention of puberty as a separate phase in life coincided with an increase in the age of marriage, a change that was brought about by changes in the production process in the age of industry.

'Youth is but a word, an empty shape with ever-changing parameters,' wrote media philosopher Pierre Bourdieu. Ever since James Dean played the rebel and young musicians started to swing their hips rhythmically in the 1950s, youth culture has increasingly concentrated on consumption. More than ever, adolescents are targeted as a distinct group by marketing campaigns. In the huge cinema complexes, those belonging to other age categories have been marginalized in favour of teenagers with lots of pocket money and little inspiration. The aseptic, hyper-aesthetic images that dominate the media can only cause frustration. The most obvious means of compensation seems to be consumerism, for the adolescent does not know how to cope with his or her desires (strongly cultivated by others) other than by spending money. Everything is done to make us believe that a certain, well-defined image relates to adolescence, and that this image can be marketed. The American artist Dan Graham has pointed out how Brooke Shields's success as the eight-year-old Lolita (thanks to Garry Gross's series of photographs *The Woman in the Child*, 1975) created a market in the United States for a new line of cosmetics, underwear and other luxury items for children. In his collection of essays *Rock My Religion* (1993) Graham linked the cult of adolescence with politics. Today, however, this sort of critical reflection is dismissed as the pointless meditations of a killjoy. An increasing number of adults in fact create an improbable profile of themselves as 'kidults', living the illusion that after an utterly boring working day they can return to adolescence. Just as Teletubbies merchandising was reclaimed by adolescents, adults now reclaim territory from the teenagers they once were themselves. With imbecile logos and by simplifying every form of the experience of reality, marketing and merchandising keep young adults

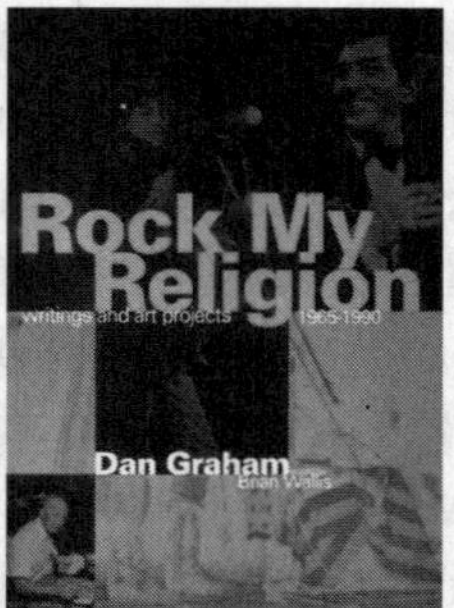

in a permanent state of infantilism. In Disneyland – and other theme parks – these days, there are as many adults on their own as there are with children.

Exemplary?

Adolescence today is not merely seen as an important stage in life; it is considered the only stage in life that can really be enjoyed. The intensity rapidly decreases but is maintained artificially for as long as possible. Adults are already referred to as 'fading adolescents'. The best part of our lives is short-lived, if we are to believe the logic of marketing. Apart from this regressive urge, there is also the fact that adults like to remember themselves as an object, i.e. as the prototype of the adolescent, but prefer to forget about themselves as a concrete subject. The adult consciousness, for that matter, declines to be transferred back to adolescence. The feeling of unrestrained freedom of expression and of movement may be the ideal to aspire to, but the accompanying feeling of physical malaise and tormenting uncertainty is erased from memory. In 1999 the film *American Beauty* was met with almost universal acclaim. Instead of presenting a critical discourse on the callous indifference with which employees are made redundant these days, the film unashamedly cashes in on the urge to live out a second puberty and, of course,

throws in a fixation on a Lolita character. A sugary
daydream is celebrated as the ultimate act of
rebellion. Jumping on the same bandwagon of
dubious nostalgia, Belgian fashion designer Martin
Margiela created a collection of models based on
dolls' clothes. These have been proportionately
enlarged (including press-studs and accessories)
to fit adults. The collection can be seen as a
parody on the role play – the utopian typecasting,
the Barbie-meets-Ken syndrome – children lose
themselves in. At the same time it alludes to the
bitter disillusionment that awaits them if they cling
to these promises of a fairy-tale future. Walter
Van Beirendonck designed a brightly coloured
baby collection in which children literary start to
twinkle like radiant angels or attractive aliens:
small, battery-powered lamps were woven into the
fabric. Van Beirendonck's conspicuous necklace,
on the other hand, was inspired by the identity tag
every (potentially doomed) soldier wears round his
or her neck.

Every type of clothing implies certain expectations. Clothes not only construct; they also
determine the wearer's experiences to a large
extent. In adolescence, an awareness awakens that
every garment inevitably sends a signal. But even
without these various packaging strategies, humans
increasingly seem to be a construct. Variations in

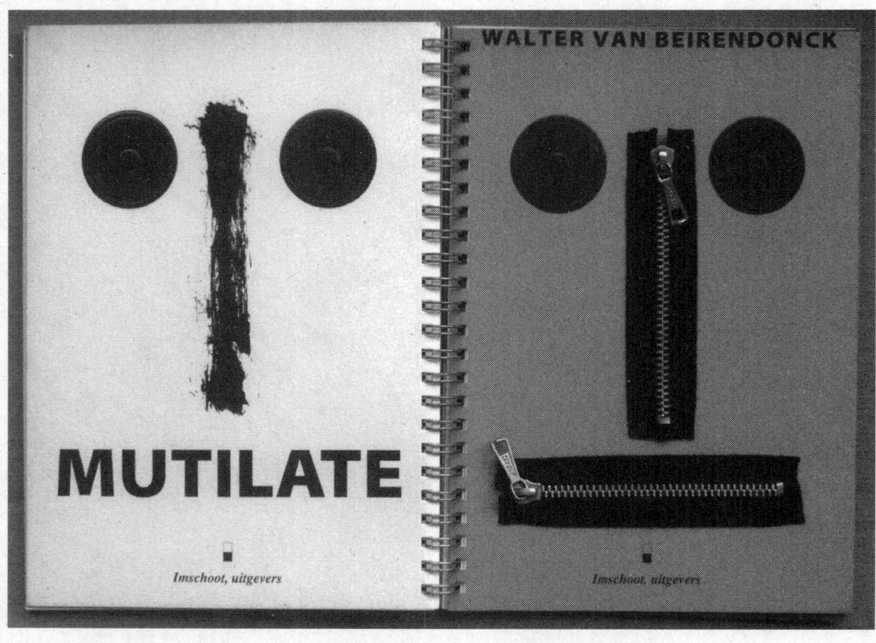

8.14 Walter Van Beirendonck & Paul Boudens, *Mutilate*, 1998

the idea of beauty concern not merely clothing, but our physical sizes as well. The controversial heroin-chic look that was fashionable a few years ago might well be interpreted as a desperate longing to withdraw into one's early or even pre-adolescent body. Dutch photographer Inez van Lamsweerde presents portraits of imaginary figures; the plastic surgery she practises is the result of digital image manipulation. *Thank you Thighmaster: Pam* looks like an eerily perfect cyborg, a mannequin that has become flesh, with perfect measurements and without pubic hair or nipples. Mike Kelley has referred to dolls as dysfunctional images of children: 'A doll is the image of a dead child, an unattainable ideal that results from the social concept of what a family should be like.' In contrast with this image, Kelley's painting *Incorrect Sexual Model: Mommy's Penis* offers a blatantly erroneous image, an image that is as phantasmagoric as it is schematic, of what a child may imagine underneath his or her mother's skirt. Or is the work a parody of the censored diagrams the child is confronted with on the classroom blackboard?

Knickers and Sniggers Comedy?

The surrealist absence of bodily fluids in the prevailing image of teenagers and in images destined for teenagers is compensated by Wim

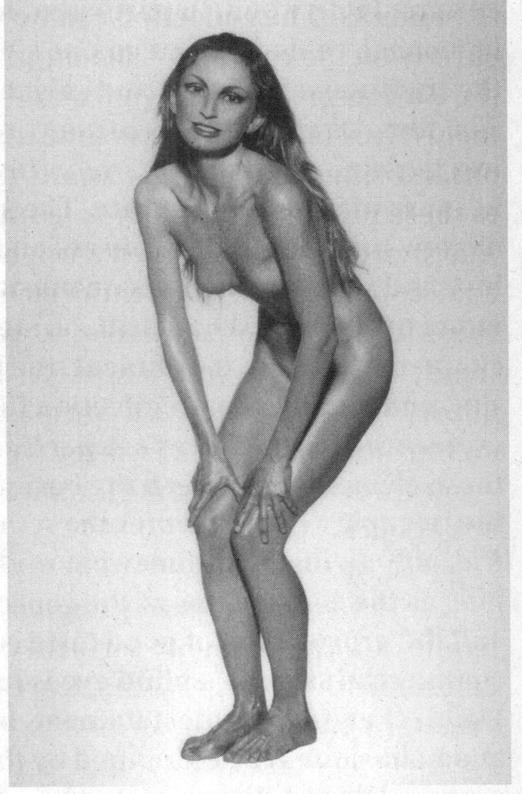

8.15 Inez van Lamsweerde, *Thank you Thighmaster: Pam*, 1993

Delvoye in his video installation *Sybille* by focusing solely on the squeezing of blackheads. At the MuHKA in Antwerp, Delvoye's dazzling installation *Cloaca* later bubbled away, a machine that mimicks the human digestive system (including the production of the final turd). Scatology, however, is not entirely banned from the official youth culture. Hollywood, in particular, has specialized in teenage comedies that go one step further than the traditional 'knickers and sniggers' comedy. The film *American Pie* is in many respects far less hypocritical than *American Beauty*, precisely because of its overt vulgarity. The small-town narrow-mindedness, the obsession with falling in love and sex, the emphasis on music, on decisive moments like the school ball, on stereotyped characters like the top student, the playboy and the slut, and, finally, the predilection for 'soft' actors: ever since George Lucas's *American Graffiti*, all these elements have been the basic ingredients of the 'teenpic'. They re-enter the set of *American Pie*, now giving off a somewhat rank smell. But as long as the conventions of the genre are respected and the provocation of good taste is limited to a few commercial stimuli, a blind eye is turned. 'When I subject myself to entertainment, my desires and behaviour are determined by the laws of the market. When I discover what I find pleasurable myself, I am the one who writes the rules,' Jean-

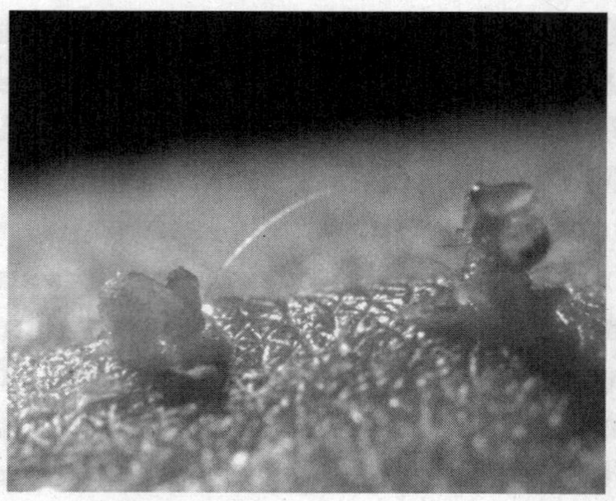

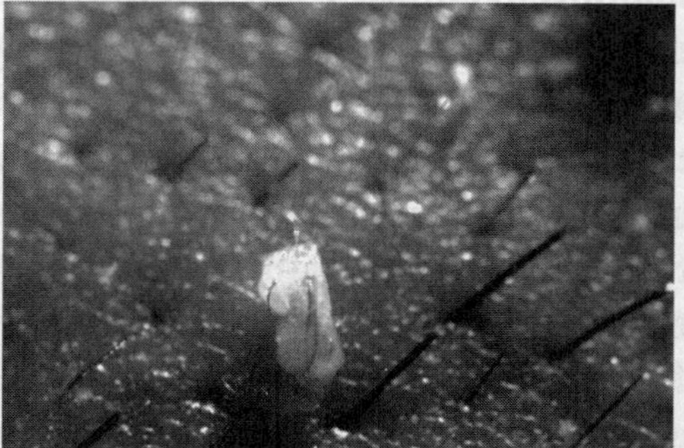

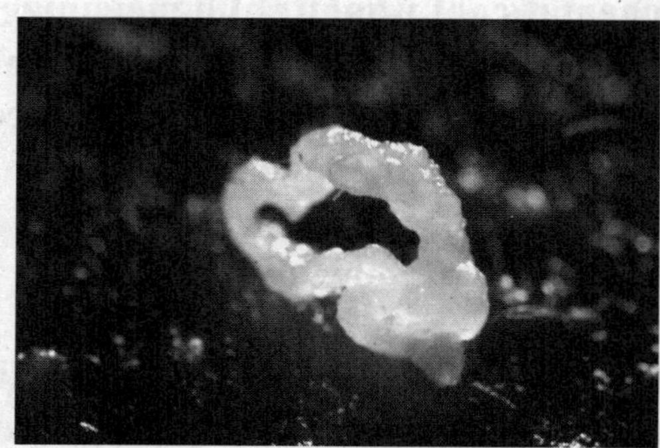

8.16 Wim Delvoye, *Sybille II*, 1999

**Charles Masséra writes in the catalogue of
Presumed Innocent ('The Becoming-Smurf of the
Northern Hemisphere – Pornoland and Happy
Market Shares Screw the User.')**

**In her video *Scrub*, Belgian photographer
Elke Boon confronts us with a naked young
woman, sitting a puddle of water, scrubbing the
floor. Yet the muddy stain grows larger and larger.
The compulsive scene suggest several layers of
meaning, but the general impression is definitely
unpleasant. Is what we see some sort of arche-
typical scene, something like a blow-up of a foetus
that turns against the placenta? An expression
of paranoid distaste the woman feels for her own
body and everything it secretes? Or simply an
emblematic image of the hopeless, Sisyphean toil
that so often is the lot of women: the home as an
endless, desperate chore. Elke Boon does not care
for labels or etiquette. The teenagers she prefers
to portray do not adopt decent poses. In her triple
evocation of *Mike*, the artist also seems to play with
female body language. According to the organizers
of *Presumed Innocent*, Stéphanie Moisdon and
Marie-Laure Bernadac, the reason why the film
Baise-Moi ('Fuck Me') caused such a fuss was
not because it showed two violent girls and their
blatant experience of sex, but because the film
was directed by a woman. Filmmaker Catherine**

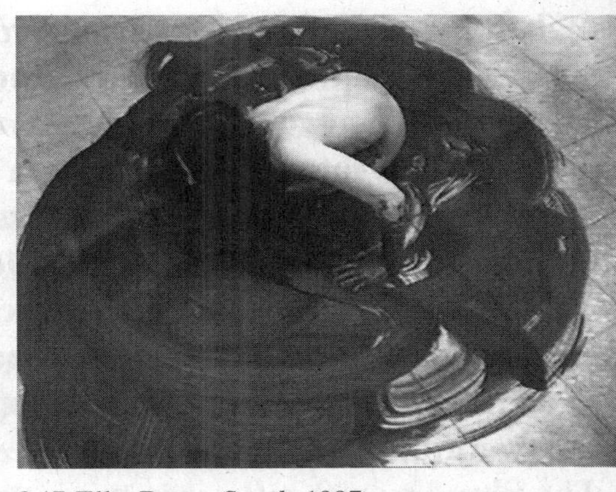

8.17 Elke Boon, *Scrub*, 1997

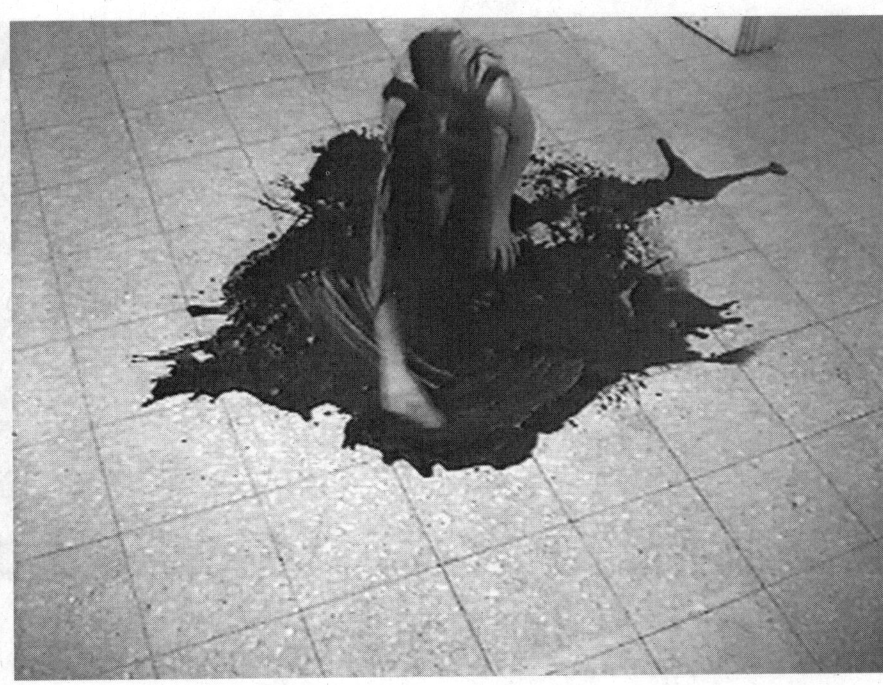

8.18 Elke Boon, *Scrub*, 1997

Breillat has presented an unabashed female view of the phenomenon of lust throughout her career, but only relatively recently has her work met with some acclaim (in 1999, for instance, a retrospective of her work was presented at the Rotterdam Film Festival). With her CD-ROMs, the Australian Linda Dement confronts the public even more intimately with her libidinous thoughts and dreams. *In My Gash* is an interactive CD-ROM that tempts the viewer into a polymorphous, perverse universe that abounds with the paraphernalia of sex and violence. Nothing in this universe is certain or well-defined: flowers unfold into the grubbiest configurations of the body and technology.

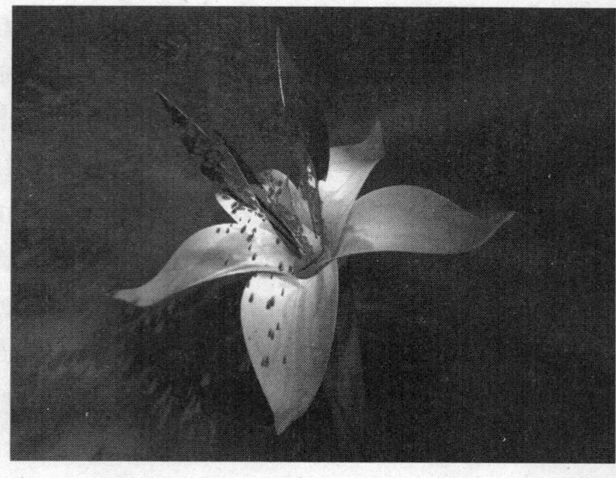

8.19 Linda Dement, *In my Gash*, 1999

Kidnapping?

It was Sigmund Freud who introduced the term 'polymorphous perverse' to describe the stage in which children are as yet unaware of their gender and in which their fantasy world has not yet been conditioned by role patterns. Though psychoanalysis has lost much of its authority (to neo-Darwinism) in the last several decades, it continues to play a role, notably because Freud was the first to consider the child as an autonomous individual, separate from moral and/or religious obsessions. Today, Freud's ideas are scarcely discussed outside film theory (and in this context

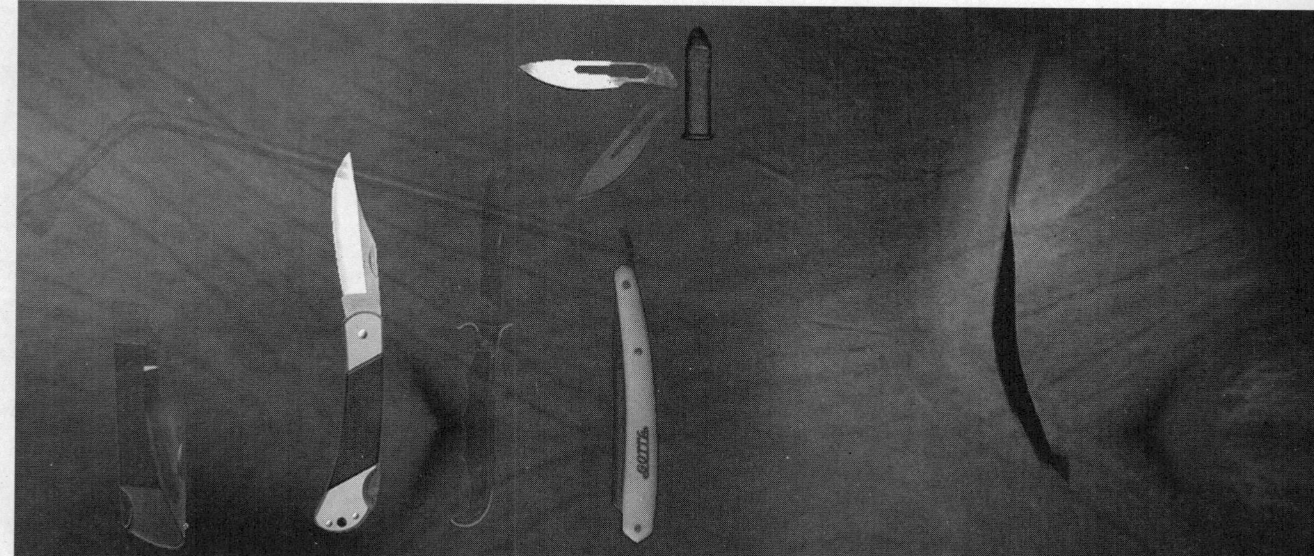

8.20 Linda Dement, *In my Gash*, 1999

they are mostly viewed through Jacques Lacan's interpretation and annotations). Freud's ideas, combined with a feminist critique of film, thus systematically undermined the legitimacy of the male perspective in commercial film culture. Terms used in this context are 'scopophilia' (compulsive watching), fetishism, and 'the Name of the Father'. The later phrase refers to the social order, the rules of conduct that are propagated by education and instilled in the mind of every citizen. One becomes a grown-up by subjecting oneself to these rules – much as an adolescent has his thick mop of hair cropped when joining the army. In Japan, social control loosens its grip on life only for a few years. Between the black-and-white school uniform at school and the black-and-white tailor-made suit at the office there are only a few years, during which Hawaii-shirts and colourful haircuts are almost obligatory. But the 'excesses' of Japanese teenagers usually take place in such an orderly and predictable manner that they cause hardly any concern to the police.

From the moment the child is born (or even before, with the aid of ultrasound devices and other instruments) the child is subjected to surveillance. With scientific tests and pedagogic paradigms, parents and teachers continually try to get a grip on the child's development. As the parents' grip

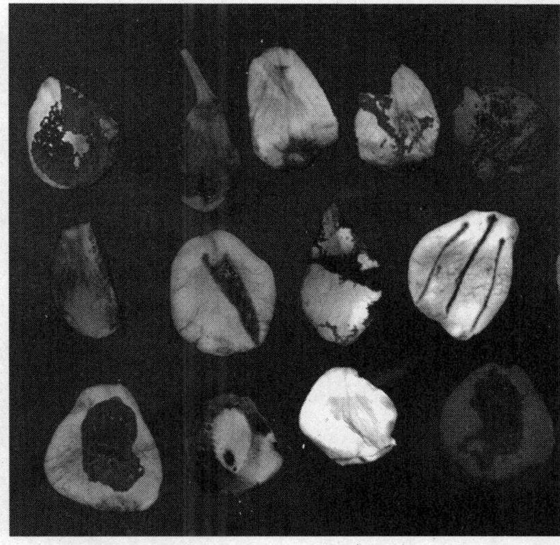

8.21 Linda Dement, *In my Gash*, 1999

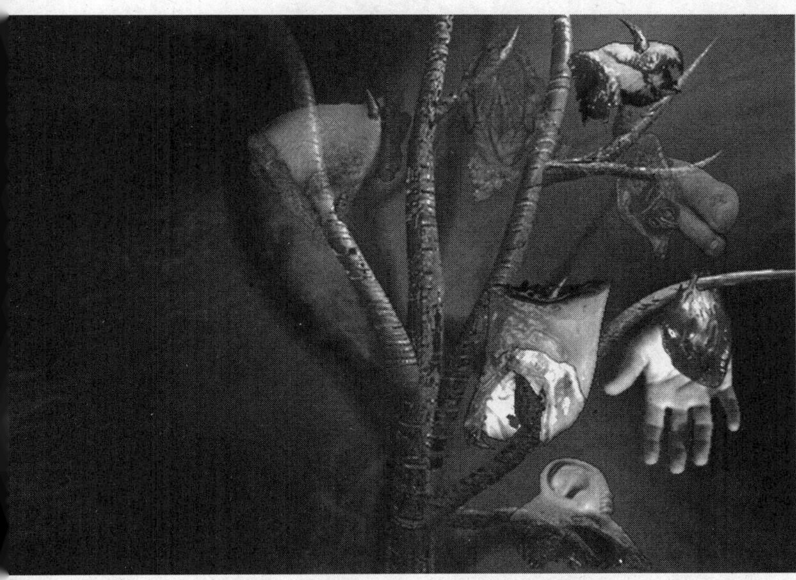

relaxes (because they desperately try to cope with the social and professional pressure they are subjected to) children are all too often dumped in front of the television (which can now be equipped with a censoring chip). Babysitters, for that matter, are monitored by video cameras. Big Brother has metamorphosed from an ogre into an esteemed confidant. When are young people still allowed to enjoy an unguarded moment, literally and – especially – figuratively? Can the young, who are targeted in so many ways, still somehow escape these controlling, conditioning eyes? Günther Förg's *Twin Sisters* cast a sceptical glance at us as soon as we position ourselves between them. They can bear only their own mirror image.

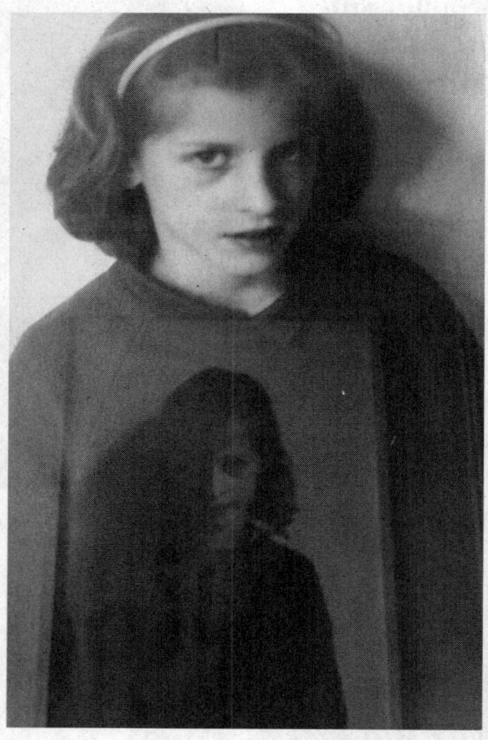
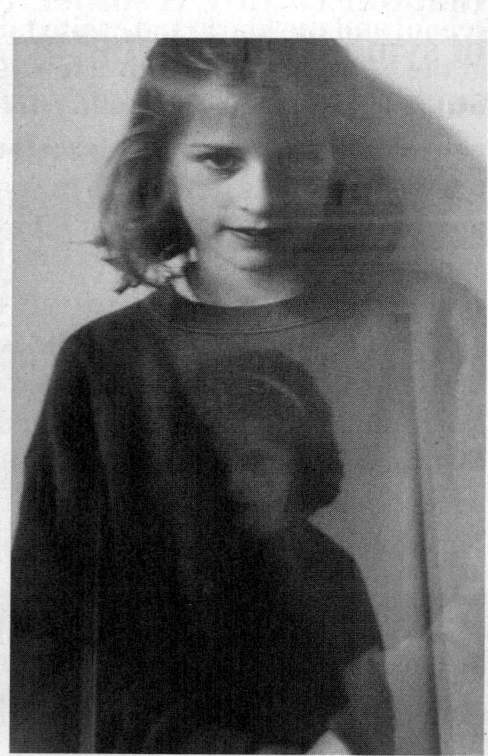

8.22 Günther Förg, *Chambres d'Amis*, 1986

Anna Tilroe: On Values and Symbols

Conversation with: Bregtje van der Haak, Chris Keulemans, Jos de Putter, Marjolein Rothman, Anna Tilroe, Annelys de Vet, Cor Wagenaar, Rutger Wolfson and Alex van de Beld.

Rutger Wolfson (RW): In Anna Tilroe's article 'Het grote gemis' ('The great absence'), published in the *NRC Handelsblad* newspaper a few weeks after the murder of Theo van Gogh, she writes that our culture is subjected to all kinds of forms of symbolism, but that in times of great tension, such as in the wake of Van Gogh's murder, symbols genuinely perceived as meaningful prove to be lacking. They can no longer be produced. We have to make do with the orange bracelets promoted by the entertainment industry, for instance. In that piece she also describes how it is that today's artists are no longer able to create these symbols. And then she calls on them to break this impasse.

From the vehement reactions to the article it became apparent that there is a great deal of reticence about symbols in artistic circles. The prevailing view is that we have had very bad experiences with symbols over the last century, and that we should therefore, in fact, refute and puncture symbols. Does this also mean that

the creation of significant symbols has become completely impossible? I thought it would be interesting to organize an exhibition around this theme and to invite an artist, a designer, an architect and a filmmaker to respond.

The big question, of course, is what these symbols should express. And because it is unfair to lay all of the responsibility for this on the artists, we felt it would be a good idea for them to debate this among themselves along with a small group of thinkers from various disciplines. In essence this think tank even serves as a kind of commissioning entity.

To this company, therefore, I pose the question: What, in this day and age, should these symbols for the Netherlands stand for?

Anna Tilroe (AT): When you think about genuinely powerful symbols, you should talk about the values they would have to represent. Yet this discussion is still taboo in artistic and intellectual circles. To talk about values means to talk about moral conscience, and that is a subject that can get you into a lot of trouble these days. We are still living – and Van Gogh's film *Submission* is an example of this – with the idea that 'anything goes'. As a result, the debate about important moral issues – such as what we find tolerable and what we no longer find so, where the freedom of the one becomes the denial of freedom for the other – is mainly being waged in the religious sector of society. This is embarrassing, and it is also alarming, because in reality it means that we, artists and thinkers, have largely relinquished the discussion about what kind of society we would like to live in and what kind of people we would like to be. I am therefore entirely in agreement with Jos de Putter when he argues, in an interview in *NRC Handelsblad*, that art, because it is interested only in its own domain, has lost its interest in the social domain.

Jos de Putter (JdP): In the gigantic confusion currently reigning in the Netherlands, things at which we used to laugh out loud are being taken seriously. Geert Wilders's political programme, for instance. It's so idiotic, so ignorant. But there seem to be no criteria by which to judge anything. In that sense, you could say that a complete absence of symbols is actually symbolic of the present situation.

Cor Wagenaar (CW): I think we should first agree that when we talk about symbols here, we are not thinking about something personal but about something collective.

JdP: The Great Absence. That, in three words, is where we've ended up as a result of the events of the beginning of this century. If you watch an evening of VPRO television programmes about education in the Netherlands, it suddenly seems we've sunk into some sort of quagmire and we don't know how to get out. Yet as a documentary maker I only feel qualified to document that void, not to fill it.

Marjolein Rothman (MR): As a visual artist I feel the same way. By showing what you actually reject, or experience as a void, you create more space, make something possible again.

RW: That is the traditional viewpoint. But that is exactly what is being called into question now. Many people point out that neo-conservative, Christian-oriented American politicians employ a very clear symbolism: the America of the 1950s, security, the family, hard work, God as Truth, that sort of thing. The American leftist elite, in response, remains silent, because it finds it abhorrent to use that brand of imagery. American sociologist Susan Neiman has noted that leftist-liberal America is paralyzed by a fear of moral kitsch. Taking a cue from Neiman, I wonder whether symbols and images could not be developed to counter, for instance, the neoconservative movement or extremist violence. So not just showing the void, but filling it too.

Bregtje van der Haak (BvdH): Koolhaas has done that. For his project for the European Union, he was commissioned by Prodi to come up with positive symbols for Europe, and that's what he did: alongside a gigantic collage for an alternative European history there was, for instance, a gorgeous flag with a multicoloured bar code. It produces new images, but I wonder whether such a commission to design symbols really can change anything about the way in which people view Europe. If there is a problem with Europe, if its citizens have no confidence in the idea and in the leaders of the European Union, I don't think the solution is simply to have an advertising agency, an artist, an architect or a designer or whoever come up with an image that is so attractive that citizens *will* start to believe in it. Every problem these days is seen as a 'communication problem', whereas the problem of Europe is far more substantive. I consider it a kind of scam. And it's a good thing that people today are no longer so easily bamboozled by clever communication. As a television programme maker, I can only relate to the absence. Doing that in the right way is difficult enough in itself.

Annelys de Vet (AdV): There's another way to look at it. Europe is a new idea, and concepts like that are so difficult to fathom that it feels like a void. Yet perhaps this void is not just an absence of

content, but also an absence of good visualization. It's likely that once the idea of Europe is properly visualized, we will be better able to relate to it in a substantive way.

BvdH: That would be great, but I'm not holding my breath.

AT: Cor Wagenaar mentioned a collective symbol. I think that we have a problem, ever since the collapse of the ideologies, in that we no longer dare to think about what it means to live in a community. In our society it is assumed that the individual lives solely for him or herself as a calculating citizen. In addition, calculating minds in the art world carefully maintain the Romantic myth of the artist as a hyper-individual, something artists themselves believe all too willingly. I think that, if we want to think again about society and the position of art within it, we have to start by critically examining this glorification of the individual and start thinking about community again.

JdP: But isn't that a question of quality, as opposed to a question of being? Isn't it the case that it is inconceivable to create something that is anything but intensely personal, and that how many people it reaches depends on the quality of that intensely personal expression? That's what I still think. When you, as an artist, set yourself the goal of reaching as many people as possible, you're treading on a slippery slope. Where is the beginning and where is the end?

AT: There I agree with you. But I wonder whether it wouldn't be an immense relief if we started distinguishing the artist as an individual from the work of art. Once the artwork is outside the studio, it is taken away, to a greater or lesser extent, from the intensely personal thinking of the artist and it starts functioning on other, more social levels. Everyone knows that, but no one in the art academies says it to young artists. So they start out with a completely false idea of themselves and of their art.

JdP: In a film I made about Europe, I talked with young architects from the former Eastern Block. They turned out to have a keen sense of the function and the effect of images in public space. Theses images were strongly oriented toward discipline. As a conceptual exercise I asked them, 'Imagine you're asked to design an European town square; what would it look like?' In any event, no statue, no gesture, they said. You have to be able to see the space.

Chris Keulemans (CK): I think an incredible number of people here need a beautiful, generous, understandable, inspirational image in their environment.

JdP: What stayed with me from making that odd little film is that people from all walks of life there do not have a sense of a great absence, but talk quite comfortably about the soul, spirituality and identity. They have a self-evident definition of those concepts. We lack that.

BvdH: Look at the European Union. In reality it is more or less a floating island. The Union exists as an economic principle, but beyond that Europeans don't know what's in it for them, because the idea has no social or ethical content. It does to outsiders. Americans know very well what Europe stands for, but from the inside people don't see it so clearly. The French want to protect their own agricultural subsidies and their own employment laws. We don't want Spanish asylum seekers in our country. To be able to call yourself a European, you have to discuss and test the history you share with one another. For that you need a European public space, in which European magazines circulate, a European news-paper is published and European television is produced. Only then can a process of European public opinion making take place. I think that this will be a far more significant issue in the future than the question of whether we as individuals are being satisfied.

AdV: But still you cannot deny that in the past, the younger generation was always a little bit better off than that of its parents. That is no longer the case, and the trend seems irreversible. This is one major reason for the 'atomization' of society. It undermines the ultimate solidarity, the solidarity between the generations. And that makes it very difficult to find symbols for what still connects us.

BvdH: That's exactly the reason we're looking for them. In this phase of capitalism, everything at the production level has to happen on a larger and larger scale, cross more and more borders. No one can explain why, but there has to be constant economic growth in order to maintain what we have. On the other hand, at the personal level, the units are getting smaller and smaller. There are 3.5 million single households in the Netherlands. The birth rate has never been so low. Very many children have no siblings, and hardly any cousins. This means that the ties that, to a significant extent, make you who you are, the ties to your roots, to your family, are getting more and more fragile. When you add the fact that work and the circulation of money and goods are organized on an increasingly large scale and entail an increasing degree of abstraction, it is clear that, for the individual, an enormous gap has developed between his or her social reality and the economic reality upon which he or she depends. You have fewer and fewer stable links with the people in your immediate environment, and at the same time you have to relate to something that is actually too abstract, too big and too scary to bear thinking about. You might say that the inside world is shrinking and the outside world just keeps getting bigger. It's hard, as an individual, to know where you stand, who you are, where you belong.

AT: You're describing a world in which the economy not only predominates, but in which no riposte can even be formulated.

CK: Nothing's stopping you from wishing.

BvdH: That wishing is the absence. It comes from the soul of the individuals who see great abstractions and experience an absence.

AT: The question is, what is being wished for?

RW: The hysteria following the murders of Fortuyn and Van Gogh made people long for archetypal Dutch values like rationality, common sense, tolerance. These seem to have been lost, to have become awkward, even though they've brought much benefit to the Netherlands over the years and given it an identity.

BvdH: With the idea of national identity, you quickly run into major misunderstandings. Very many children in the Netherlands, for example, don't recognize themselves in the national history presented to them in school. It is a historiography that barely jives, if at all, with the new reality in which they – and therefore we – live, a reality in which diverse cultures and histories exist side by side. One single narrative as general history no longer suffices. There are calls for it at the moment, but it's no solution.

AdV: We're looking for a new narrative that gives us the sense of being connected to one another by invisible wires. We've lost that story. The point is, what is the new big story that we want to write?

CW: I think it's always been a vision of the future.

JdP: A wish for a common future.

AdV: In our old passport the history of the Netherlands was pictured. In the new passport there are only numbers. Bank notes and postage stamps used to feature Dutch symbols. All that has dis-

appeared and nothing has come along to take
its place.

BvdH: That is typical of a government that is really
concerned only with streamlining the economy,
social interaction and the landscape in such a way
that everything can be employed as efficiently as
possible in the economic system. The state
addresses the citizen as a calculating citizen.
All regulations are based on the assumption that
people will choose the options that are most
economically advantageous. The regulatory regime
is predicated on this. Meanwhile, in practice,
people perform an incredible amount of unpaid
work; they take care of people who need help and
they bend over backwards to organize themselves
and combine all sorts of tasks, half of which are not
acknowledged. Millions of Dutch people work as
volunteers: unpaid, invisible. This demonstrates
that people are far better, morally speaking, than
the state always assumes. That itself is a reason to
be hopeful. Only you have to develop a kind of
language that makes you think, 'I want to be part
of that; it's good to do something for someone else,
for nothing.' That makes it, shall we say, hip to do
volunteer work.

AT: There's our first value: Altruism.

CW: You could also mention Curiosity, in the sense of seeking out knowledge, showing interest. There is far too little curiosity about other people, other lifestyles, other cultures. In that, the media take the lead.

RW: If you link Altruism with the economy, you could link Curiosity to the media. The media are always accused of not wanting to look beyond the clichés.

AdV: Contact. There is a very big gap between people and things, between messenger and message. In personal contact you immediately get involved; you're alert; you're expected to take responsibility.

CW: Contact as the polar opposite of the loss of social connections.

RW: I'm noticing that I find it almost scary to say what matters to me.

AT: Yes, some ideas evoke a certain embarrassment because they have an almost sentimental connotation. An idea like Attention, even though that is certainly something there is a great absence of.

AdV: Giving attention or getting attention?

RW: Both. Modern society is about speed, transience; attention has no place in that, because attention is something that slows things down. Attention is also part of Contact. I heard that a third of the population today are members of dating services!

MR: To be honest, as an artist I don't feel inspired by this list of concepts. Representing Altruism: how do you do that without it becoming a morality play? I'm more looking for a way to describe the absence through a work of art, to encircle the void, as it were.

AT: But that's what art has been doing for so long. Wagner complained about the emptiness he felt all around him. He thought he could fill that absence by coming up with new myths.

MR: But stuffing a series of values into art ... In the worst-case scenario that becomes propaganda.

RW: Why would an image of altruism be propaganda?

MR: It isn't propaganda per se, but I do think it has to be a personal image as well. Then it can exude pride or power, but at the same time it's about other things as well.

RW: In reality you're arguing for the autonomy of art. Yet it is precisely that autonomy that has driven art into social isolation.

BvdH: I think that this so-called autonomy of art is the reason many people have become disappointed by art and think, 'That's not for me; those visual artists are just chasing their own tails. In the end their art just winds up in hoity-toity galleries and in the houses of the rich upper class. It's got nothing to do with me.'

MR: That's a totally distorted picture. The fact that I don't paint a junkie with a needle in his arm doesn't mean that my work is not about society. There's a lot of art being made right now that is critical of society.

AT: Yes, but critical based on a negativity: 'it's all wrong'. I admit I myself am a major representative of that, but if you take the position of art in society as a gauge of a level of culture, you also have to ask yourself how it is that art has become so alienated from the social domain. Maybe you have to conclude that this negative critique has ceased to function as a 'healing factor'! In that case, it might be an interesting conceptual exercise to try something different, for instance based on a non-apocalyptic vision of the future.

CW: If we're looking for a symbol that expresses a wish for the future, then I would really want personal freedom back. Now the state, all sorts of institutions and the media are meddling into our private lives without restraint. There are cameras everywhere; all e-mails are checked; all internet providers are required to record every page you've ever visited. And no one protests. Big Brother has become everybody's hero. Until recently that was inconceivable; people thought it was totalitarian. Now that personal freedom has been relinquished and everyone accepts that there is absolutely no respect anymore for personal privacy.

MR: The right to personal privacy has to do with Dignity.

AT: The nice thing about Dignity is that each person has to define it entirely on his or her own, otherwise you're immediately shown up as a fraud. If it isn't yours, if it isn't unique to you, then it is not dignified. So you need Courage as well.

AdV: Aristotle said that Courage is the very first quality, because Courage brings out all the other qualities.

MR: I find Courage a rather scary concept, something that fits in with the current mood of 'actions speak louder than words'.

CK: But that's exactly what we're doing: hijacking these words from the people who control them – the Christians, the New Right, etc. They've appropriated these ideas and we've been left empty-handed. So we have to take them back and try to give them a new meaning: our meaning. But it might be funny to include one word, among all the lovely concepts we're putting together, that would work as a kind of torpedo.

MR: Civil disobedience.

CK: Ever since the murder of Van Gogh I'm constantly getting into discussions about where we go from here and what one should inevitably conclude as a right-thinking person. Since that time I also engage strangers in conversation; I behave properly in public; I don't insult people; I don't always say what's on my mind; I respect you; I'm curious about how you live – in other words I'm becoming a sort of model citizen. And that's exactly what I never wanted to turn into!

AdV: The weird thing is that being a model citizen is now almost a form of resistance, whereas 10 years ago it was the worst thing you could be.

MR: So, Resistance.

RW: We now have Altruism, Curiosity, Contact, Dignity, Courage, Resistance. For art and for the Netherlands.

CK: It is a weird move to ask an artist, of all people, to step over his or her own shadow and create a symbol for the lost community of the Netherlands. It's even weirder that this is being initiated from an artistic place without having been requested by the government or the citizenry. It really is, from all points of view, a completely impossible challenge, and that is exactly what I find appealing about it. The whole project flies in the face of everything one thinks about art and the community. Actually it's impossible, and in that sense it is perfect for art, because art always thinks about things that are impossible.

On Symbols (second conversation)

RW: We've now put together concepts that we think stand for important values and for which the artists who take part in this project will have to provide a symbolic form. Chris Keulemans has noted that in doing so we want to give new meaning to values that have become threadbare or that have been 'stolen' from us by religious fundamentalists and neoconservatives. The question now is, whom do we mean by this 'we' who have been robbed?

AT: The cultural elite. When the cultural elite does not assume its responsibility to society, things go wrong. A British psychiatrist wrote that in a book about his work with the underclass in Britain. In it he points out that this underclass is not capable of taking responsibility for its actions and argues that the cultural elite is unable to propose or do anything in response. It is as though it were suffering from a great absence, a void that paralyzes. And that trickles down to the lowest levels of society.

CK: The so-called 'crisis of the intellectuals'.

JdP: I posed several questions about this to the Hungarian writer György Konrád. He said there is no such thing as *the* intellectuals nor any such thing as *the* cultural elite.

AT: To me the cultural elite is composed of people who are involved or feel involved in the shaping of philosophical and societal ideas and their discussion.

CK: Surely the whole controversy in the Netherlands, the discussions back and forth about Islam and democracy, is all being waged and funded by a cultural elite?

JdP: That's the question. Is Geert Wilders an intellectual?

CW: The criterion is participation in the discourse, not so much what someone says or what viewpoint he or she espouses.

JdP: These days, public opinion determines the shaping and expression of ideas. The media take their cues from it, and this is how we've ended up on this sliding scale. It is precisely the opposite of the way in which Thomas Mann, for instance, defined his viewpoints during the First World War. He took only himself as a criterion. Where we sit, at the other end of the spectrum, the viewer or the newspaper reader are used as touchstones for

how problems should be described. This viewer
or reader doesn't actually exist, of course; he's
constructed out of numbers and surveys. And the
conclusion is that the discussion shouldn't be too
difficult. That is just as damaging as Mann's intellectual isolation.

CW: Today people first check out whether there is
a market for an idea.

CK: Everyone is in some network or other. The
figure of the thinker in isolation no longer exists.

CW: Perhaps it does, but we'll never know.

Alex van de Beld (AvdB): As an answer to the
media bombardment you should in fact dare to
make personal choices, even if they are temporary
choices. To free up ideas it is sometimes necessary
to firmly shut the door behind you. That makes
this project so difficult and perhaps even over-
ambitious.

AT: That's why it is imperative to emphasize a
clear perspective. Based on argumentation and
in a debate.

RW: That perspective could be what art can
contribute to society in the realm of ideas.

MR: An artist is not a politician. Surely art should not prostrate itself before the public and say, 'Look, I'm not doing just anything; I'm also doing it for you; it's in your interest too.' I think people should try harder to understand the language of art. Then they would see that art is in fact relevant and says something about the world in all sorts of ways..

AT: But the issue here is not whether art should or should not prove its function. That discussion has been waged so often before, and always in the same familiar terms. It advances nothing. What we're trying to do now is find an angle that can bring art in a closer relationship with the social domain without it ceasing to be art.

MR: But the reason art has become so powerless also has to do with the evolution of society. It's very difficult for an artist to adopt a position in the midst of a society as fragmented as ours. It's not that I'm against the idea of art being functional. In fact what I mean is that it is, but today's society refuses or is unable to see that.

AT: Yes, but there's another party involved: the art world itself. It has systematically failed to communicate clearly what it is doing in the first place. Art theorists have come up with an incestuous language through which they keep their entire discourse

closed. Criticism ultimately slides off of it, or leads to ridicule and excommunication. The result is that outside the art world, and to a significant extent inside the art world itself, no one is interested in the visual art discourse.

JdP: I understand Marjolein Rothman's problem quite well: there is no intermediate zone between the social domain and the artistic domain. That's in fact why the step we are taking is so risky. The start of the solution might well lie in words like dynamism and flexibility. You have to use rubber to make something that looks like steel. Values have a hard quality, like a thing, whereas an attitude, for instance, moves. When you're thinking of attitudes, something like the concept of Compassion comes up. Compassion is not a thing in itself, but an attitude toward something that is real, that exists in the here and now. You can define that attitude in all sorts of ways, including an artistic way. This is gentler and therefore more open to a flexibilization of the problem.

AvdB: Making the values fluid, I think, is a lovely metaphor, especially because otherwise they can too easily be lumped with the Christians virtues.

CK: On the other hand, we want to re-appropriate certain 'hard' words like Courage and Resistance.

That's why I think the commission brief to the artists has to be formulated as specifically as possible. What they do with it from then on is up to them.

RW: I agree with you there. We have to try to provide an impulse for a discussion about why there is so much aversion to values.

AT: I also think that the age we're living in demands new meanings for words like Courage and Resistance. Western society is increasingly ruled by fear. That raises the question of how you deal with that fear. Do you deal with it flexibly? What does courage consist of? The same is true of resistance. If you have the feeling that the globalization of the world is washing over you, that a huge capitalist system totally controls your life, you can adopt a position of resistance to that. Or at least go in search of one.

JdP: Perhaps it would be useful to think out loud again about classical concepts that have long been associated with art: Good and Beauty. We agree that Beauty has detached itself from the world and then shut itself up in its own discourse. It has become separate from the Good that is something to do with society, something for the people. Beauty should therefore be redefined so that it has to do with the world again. This is not the first time such an attempt has been made. Are there examples in history in which it turned out well?

AT: In the first three decades of the twentieth century, modernism was closely allied to Socialism. Art expressed the ideals of politics. The Bauhaus, of course, is an explicit example of that. And of the bankruptcy of that fusion, because ultimately it was savagely killed off. Idealistic politics and idealistic art had the same source: the idea of a world that could be created. That's the start. If you look around Eastern Europe now, you see that here and there architecture expressed those ideals in an extraordinarily impressive way.

CW: There's also an element of utopia. Last time we talked about the sense of a common future. The utopian atmosphere of the Reconstruction following the Second World War contained that sense.

CK: It's all too often been the case that artists commit themselves to a utopia and that it then reveals itself to be the total opposite. But there are certainly examples of art serving as an engine of the imagination of a new idea without committing itself to a party or ideology. A writer like Cortázar, for instance, in the 1970s in South America. Or the filmmakers in Italy after the Second World War, when the country was still extricating itself from the Fascist era. Those artists were very left-wing, but they did not submit completely to politics. The point is that you should not think that art has to last an eternity and represent a utopia. The point is that a gesture is being made that works for 10, 20 years, on the way to the next society.

AvdB: In my firm we're thinking about the word *nutopie* [a Dutch portmanteau word combining *nut*, 'usefulness', and *utopie*, 'utopia']. This is a new word that is not about eternity but about immediate usability. The trap is then that you want to stick on a symbol onto it from the outside. That doesn't work anymore. It did in the 1920s. Back then it was possible to work with concepts like modern, open and healthy. Since then, I think, we've forgot the ways in which we might make an impact. That's why I find what Peter Sloterdijk writes in *Spheres* interesting. I'll put it in my own words: don't let yourself be told what's good for you by the powers that be; set your own agenda for action, based on the here and now. From there all you can do is hope that these actions will have an effect on the community.

AT: I still detect some wishing in that. What would you like that community to become? Or human beings as individuals? You can't pose that question anymore, because to think in terms of a noble human being or, like the Socialists, a new human being has become impossible. Yet doesn't that create a void, if we are no longer allowed or able to think about what we would like to be?

JdP: The position that is formulated will always be an individual one anyway. Briefly put, this is 'I doubt, therefore I am.' Any other position today makes you part of the media circus and subjects you to norms and values imposed upon you from the outside.

AT: You can also elevate doubt to the level of something you share with others. Incidentally I have the feeling that doubt commands little respect right now as an intellectual attitude.

CK: It's a question of the ability to live with other people. This doesn't work with harnessed stipulations. In that sense, doubt is a way to start a dialogue. Doubt as a readiness to make a position debatable.

MR: May I pose a very practical question? Is the idea that, for the artists, the formulation and the brief be contained in the execution of the project?

RW: The idea behind this whole project was that you actually cannot ask the artist to define the values we need. But you can ask a club of thinkers to serve as a commissioning entity and identify the values and then submit them to the artist.

AdV: What I really want to know is what you all mean by symbols.

AT: You could call a symbol a very powerful image that represents a value that is perceived as important by many people.

AdV: That means it has to be understood by many people, to provide clarity.

AT: Yes. But a good symbol also has many layers; otherwise it's a code, like in advertising. Codes are one-dimensional; they only serve to identify.

AdV: Does a symbol have to be monumental?

RW: Take the American flag; that's surely monumental. But you probably couldn't create something like that. To make that you need a lot of 'common' history.

AdV: What you're saying is the symbolic effect comes through use, not through the design. Even if the American flag had looked different, it would have had the same effect.

CW: Yes, but it does stand for an idea about a nation.

RW: There are also images that became symbols in hindsight. A symbol for the atrocities of war, for instance.

AT: But then they are not works of art. The destruction of the Twin Towers has become a symbol, but it is not a work of art. In that I continue to disagree with Stockhausen. There is no artistic thinking behind it to lend it meaning. The meaning was only added later, by the media. Nor is that orange bracelet everyone was supposed to wear after Theo van Gogh's murder a work of art. It has the fragility and the short life of a hype.

RW: A good symbol is legible, understandable and shared. This is at odds with artworks that require a lot of time and attention – which is not to say that there aren't examples of very good artworks that contain all that and yet communicate easily, right?

MR: You can make something that is complicated yet seductive in an appealing way.

RW: I hope that this ultimately produces symbols that other people can adopt and use easily.

AdV: A sort of gift to society?

RW: Perhaps, yes. Perhaps this will produce something that is a gift to the Netherlands.

Guus Beumer studied social sciences and has a broad experience within the field of art and culture. Until 2004 he was co-director (art director) of Orson + Bodil and SO by Alexander van Slobbe. Beumer writes regularly on art and design, for instance in periodicals including *Metropolis M* and *Elsevier*. Beumer is director of the Maastricht branch of the Netherlands Architecture Institute and director of Marres, Centre for Contemporary Culture in Maastricht.

Cornel Bierens trained as a psychologist and artist and works as a critic and artist. For many years he wrote about art for the Dutch newspaper *NRC Handelsblad*. He is the author of the successful children's novel on art *Schildpad met Roos en Mes* ('Tortoise with Rose and Knife', 2002).

Valentijn Byvanck is director of the Zeeuws Museum in Middelburg. He studied cultural history in Utrecht and New York and was a fellow of the Smithsonian Institution and the Metropolitan Museum of Art. From 1999 to 2002 he worked for Witte de With, Center for Contemporary Art in Rotterdam. His writings published in 2007 include *Kunst moet je voelen* ('Art is Something You Have to Feel'), a pamphlet on a new museum culture.

Edwin Carels is a film maker and curator who is especially interested in the relationship between the visual arts and film, video and photography. He works for the International Film Festival Rotterdam, the Museum of Contemporary Art in Antwerp and as a freelancer. Besides curating, Carels also teaches and writes about art, film, animation and media archaeology.

Chris Darke is a London-based writer and film critic. He has contributed to *Film Comment, The Independent, Sight and Sound, Trafic* and *Cahiers du cinéma*. He is the author of *Light Readings: Film Criticism and Screen Art*, a monograph on Godard's *Alphaville* and *Cannes: Inside the World's Premier Film Festival* (with Kieron Corless).

Bregtje van der Haak is a documentary film maker, writer and organizer. Since 1998 she has been working with Dutch public broadcaster VPRO and directed over 20 international documentaries on social, cultural and political subjects. Her documentaries include *Lagos Wide & Close* (2004), about self-organization in Nigeria, in collaboration with Rem Koolhaas and the Harvard Project on the City; *Saudi Solutions* (2006) about career women in Saudi Arabia and *Satellite Queens* (2007) about a popular pan-Arab talk show.

Bas Heijne is a writer and essayist (including for the newspaper *NRC Handelsblad*) on cultural-political topics. He has won many prizes for his critical essays (including the Henriette Roland Holst Prize and a nomination for the AKO Prize). Publications include *De wijde wereld* ('The Wide World', 2000), *De werkelijkheid* ('Reality', 2004) and *Onredelijkheid* ('Unreasonableness', 2007).

Anna Tilroe is an art critic (for the newspapers *de Volkskrant* and *NRC Handelsblad* as well as many other art journals), lecturer (worldwide) and author of several books on art: *De blauwe gitaar. Essays over de hedendaagse kunst* ('The Blue Guitar: Essays on Contemporary Art', 1990); *De huid van de kameleon* ('The Skin of the Chameleon', 1996) and *Het blinkende stof. Op zoek naar een visioen* ('The Glittering Dust: In Search of a Vision', 2002). She is the artistic director of *Sonsbeek 10*, 2008 (an international open-air exhibition in Sonsbeek, near Arhem, in the Netherlands).

Rutger Wolfson studied art and culture at Erasmus University in Rotterdam. From 1996 to 1999 he was curator of Witte de With, Center for Contemporary Art in Rotterdam and organized exhibitions as well as (co-) producing music, theatre and dance projects. He became director of the Stichting Beeldende Kunst Middelburg (De Vleeshal) in 2000. He organized several renowned exhibitions about the boundaries of the visual arts and other art forms. Wolfson compiled and introduced the anthologies *Kunst in crisis* ('Art in Crisis', 2003) and *Nieuwe symbolen voor Nederland* ('New Symbols for the Netherlands', 2005). In August 2007, Wolfson published the essay *Het museum als plek voor ideeën* ('The Museum as a Space for Ideas'). In September 2007 he was appointed director of the International Film Festival Rotterdam, part-time, in parallel with his function as director of De Vleeshal. After the festival in early 2008, Wolfson was asked to stay on full-time as its director, and he ended his work in Middelburg in the summer of 2008.

Discussion Participants:
Alex van de Beld is an architect and co-director of the Onix architecture firm in Groningen.
Chris Keulemans is a writer and journalist.
Jos de Putter is a film maker.
Marjolein Rothman is a visual artist and painter.
Annelys de Vet is a graphic designer. Her work operates between graphic journalism, copy freedom and language.
Cor Wagenaar is a historian and writes regularly on architecture history.

<u>Rutger Wolfson: This is the Flow. The Museum as a
Space for Ideas</u>
1.1 São Paulo at night, 2005, photo: Rutger Wolfson
1.2 Ana Maria Tavares, *Relax'o'visions*, Museu
Brasileiro da Escultura, 1998, Galeria Vermelho,
São Paulo, photo: Rubens Mano
1.3 Ana Maria Tavares, *Middelburg Airport
Lounge with Parede Niemeyer* and *Numinoso*,
De Vleeshal and De Kabinetten van De Vleeshal,
August–November 2001, photo: Leo van Kampen
1.4 Mall of America, 2006, photo: Rutger Wolfson
– John Maxwell Coetzee, *Youth*, book cover, 2002
– Bret Easton Ellis, *Glamorama*, book cover, 1998
1.5 Cameron Rudd, *West Witches*, 2003
1.6-1.7 *We'll slide down the surface of things…*,
installation views, De Vleeshal, September–
November 2002, photo's: Leo van Kampen

   The exhibition *We'll slide down the surface
of things…*, showed paintings by Frank Bauer,
Arnout Killian and Glen Rubsamen. The artists'
works were exhibited in a pavilion specially
designed for the exhibition by Herman Verkerk
in the hypermodernist style so typical of recent
museum architecture. The contrast between the
slick, brightly lit pavilion and De Vleeshal's dark,
Gothic architecture amplified the exhibition's
artificiality.
1.8 Arnout Killian, *Catwalk*, 2002
1.9-1.10 Geert Mul, *Tokyo FX*, 1995
1.11 Germaine Kruip, *Counter Composition*,
installation views, De Vleeshal, September–
December 2006, photo's: Leo van Kampen
– Advertisements for 'Higher' by Christian Dior
and 'Truth' by Calvin Klein, 2002, photo's:
Rutger Wolfson
1.12-1.13 *This is the Flow*, The Girl Skateboard
Company, Gyz la Rivière and Robert Rosenau,
De Vleeshal, June–August 2001, photo's:
Frank Hanswijk
1.14 *Alexandra Ranner & Cameron Rudd*, installation view, De Vleeshal, January–March 2005,
photo: Leo van Kampen
1.15 *De Werkelijkheid*, Janice McNab, De
Vleeshal, June–September 2004, photo: Leo
van Kampen

   The exhibition brought together the work of
artists who wish to reconnect our consciousness
with reality: Rita McBride, Fischli & Weiss, Michel
Houellebecq, Krijn de Koning, Germaine Kruip,
Janice McNab, Sharon Lockhart and Wolfgang
Tillmans. The exhibition was curated by Tanja
Elstgeest and Rutger Wolfson in association with
Bas Heijne. Other examples are the scale models of
abandoned interiors by Alexandra Ranner and the
paintings of derelict urban landscapes by Cameron
Rudd. Their work shows us a reality pregnant with
longing *and* emptiness. They exhibited together
at De Vleeshal, January–March 2005 (see 1.14).
1.16 Live performance by The Melvins, De
Vleeshal, November 12, 2004, photo: Lex de
Meester

   The Melvins are a highly respected rock/
metal band formed in the early 1980s, with Buzz

Osborne and Dale Crover as core members. They
performed live at De Vleeshal on 12 November
2004, playing soundtracks for three films by
Cameron Jamie: *Kranky Klaus*, *Spook House*
and *BB*.
1.17 Lucas van der Velden, *Meta_Epics*, 2007,
© 2007 Telcosystems
1.18 Toby Paterson, *Broken Arabesque*, installation view, De Vleeshal and De Kabinetten van
De Vleeshal, April–June 2006, photo: Leo
van Kampen
1.19-1.20 Chris Cunningham, *flex*, installation view,
De Vleeshal, March–May 2001, courtesy Anthony
d'Offay Gallery, London, photo: Leo van Kampen
1.21 *De Werkelijkheid*, installation view, De
Vleeshal, June–September 2004, photo: Leo
van Kampen
1.22 *Higher Truth No.5*, installation view,
De Vleeshal, January–March 2002, photo:
Johannes Schwartz
1.23-1.24 *Higher Truth No.5*, installation views,
De Vleeshal, January–March 2002, photo's: Leo
van Kampen
1.25 *Higher Truth No.5*, installation view,
De Vleeshal, January–March 2002, photo:
Johannes Schwartz
1.26 Geert Mul, *Generating Live*, installation
view, with the cooperation of Jochem Paap, Koot
and Lucas van der Velden, De Vleeshal and De
Kabinetten van De Vleeshal, August 2000
1.27 Hendrik-Jan Hunneman, *Kiss & Go*,
De Vleeshal, May–August 2002, photo: Leo
van Kampen
1.28 Tomas Saraceno, *Air-Port-City*, De Vleeshal,
April–June 2007, photo: Leo van Kampen
1.29-1.30 *Entree, Een choreografie voor het
publiek*, Krisztina de Châtel in association with
Birthe Leemeijer, installation views, De Vleeshal,
January–April 2006, photo's: Lex de Meester
1.31 Live performance by The Melvins,
De Vleeshal, November 12, 2004, photo: Lex
de Meester
1.32 *Pan Sonic*, Ilpo Vaisanen, Rutger Wolfson and
Mika Vainio, April 2005, photo: Bianca Runge
1.33 *Pan Sonic*, Mika Vainio and Ilpo Vaisanen,
April–June 2005, photo: Leo van Kampen
1.34 *This is the Flow*, The Girl Skateboard
Company, Gyz la Rivière and Robert Rosenau,
De Vleeshal, June–August 2001, photo: Frank
Hanswijk
1.35 *This is the Flow*, The Girl Skateboard
Company, Gyz la Rivière and Robert Rosenau,
De Vleeshal, June–August 2001, photo: Leo
van Kampen
1.36 *This is the Flow*, The Girl Skateboard
Company, Gyz la Rivière and Robert Rosenau,
De Vleeshal, June–August 2001, photo: Frank
Hanswijk
1.37 *AquaStaete*, Zeelenberg Architecture,
© Copyright Zeelenberg Architectuur
1.38-1.39 *Noordzee Residence 'De Banjaard'*,
Zeelenberg Architecture, © Copyright
Zeelenberg Architectuur

1.40 *Cape Helius*, Zeelenberg Architecture,
© Copyright Zeelenberg Architectuur
1.41 *The Leisure Society*, Tim Stoner and
Zeelenberg Architecture, February–March 2001,
photo: Leo van Kampen

The Leisure Society was organized in association with the television series *DNW Rooksignalen uit de Nieuwe Wereld* ('Smoke Signals from the New World'), curated by Bregtje van der Haak and Rutger Wolfson. Wolfson's text from page 36, 2nd paragraph, to page 38, 1st paragraph is largely borrowed from the text that Bregtje van der Haak wrote for the exhibition's invitation card.
1.42 Krijn de Koning, *Beeld voor De Vleeshal*,
De Vleeshal and De Kabinetten van De Vleeshal,
May–July 2000, photo: Leo van Kampen
1.43 *Terug van School*, installation view, De
Vleeshal and De Kabinetten van De Vleeshal,
September–October 2000, photo: Leo van
Kampen

Terug van School ('Back from School') was curated by Edwin Carels with works by Walter Van Beirendonck, Elke Boon, Patrick Van Caeckenbergh, Luc Compernol, Wim Delvoye, Linde Dement, Danny Devos, Günther Förg, Mike Kelley, Inez van Lamsweerde, Sharon Lockhart, Martin Margiela, Claes Oldenburg, Koen Theys and Marijke van Warmerdam.
1.44 *Terug van School*, Marijke Van Warmerdam and Duane Hanson, installation view, De Vleeshal and De Kabinetten van De Vleeshal, September–October 2000, photo: Leo van Kampen
1.45-1.46 *Safe Haven... or the Aesthetics of Safety*, installation view, De Vleeshal, De Vleeshal, January– March 2003, photo: Leo van Kampen

Safe Haven... or the Aesthetics of Safety was curated by Guus Beumer and De Vleeshal in association with the Netherlands Architecture Institute, Rotterdam and Premsela – Dutch Design Foundation, Amsterdam.
1.47 Annelys de Vet, *Concentratie* ('Concentration'), 2005
1.48 Annelys de Vet, *Contact*, 2005
1.49 Annelys de Vet, *Herinnering* ('Memory'), 2005
1.50 Annelys de Vet, *Moed* ('Courage'), 2005
1.51 Marjolein Rothman, *Roses*, 2005, in: *Nieuwe symbolen voor Nederland*, 2005, installation view, De Vleeshal, September 2005–November 2005, photo: Leo van Kampen
1.52 *Nieuwe symbolen voor Nederland* ('New Symbols for the Netherlands'), installation view Alex van de Beld, De Vleeshal, September 2005–November 2005, photo: Leo van Kampen

The exhibition *Nieuwe symbolen voor Nederland* presented work by: Ahmed Aynan, Alex van de Beld, Peter Delpeut, Jos de Putter, Marjolein Rothman and Annelys de Vet.
1.53 Annelys de Vet, *Twijfel* ('Doubt'), 2005
1.57-1.58 Cameron Jamie, *Spook House*, video stills, 2003
1.59 *Pan Sonic Live*, performance, De Vleeshal, April 9, 2005, photo: Lex de Meester

Anna Tilroe: The Promise
2.1 Schiphol Airport Amsterdam, 2005, photo: David Bennewith
2.2 Ana Maria Tavares, *Relax'o'visions*, Museu Brasileiro da Escultura, 1998, Galeria Vermelho, São Paulo, photo: Rubens Mano
2.3-2.5 Ana Maria Tavares, *Middelburg Airport Lounge with Parede Niemeyer*, De Vleeshal and De Kabinetten van De Vleeshal, August–October 2001, photo: Leo van Kampen
2.6 From LAX Airport, 2001, photo: Robert Rosenau

Cornel Bierens: We'll slide down
3.1 Edward Holub, *Woman Getting Lips Painted*, © Edward Holub/CORBIS
3.2 Frank Bauer, *Sasha*, 1999
3.3 *We'll slide down the surface of things...*, installation view, De Vleeshal, September–November 2002, photo: Leo van Kampen
3.4 Frank Bauer, *Sasha in New York*, 2002
3.5 Frank Bauer, *The Grannies*, 2002
3.6 Frank Bauer, *After Hour*, 2000
3.7 Frank Bauer, *After Hour*, 2000
3.8 Frank Bauer, *After Hour*, 2000
3.9 Frank Bauer, *After Hour (Tussi Deluxe)*, 2000
3.10 A painting by Arnout Killian used as a fashion prop and photoshopped pink for interior design book *Dutch View* by Jan des Bouvrie. *Eigen Huis & Interieur*, VNU Tijdschriften, 2001
3.11 Arnout Killian, *Bleu Beauty Mask*, 2001
3.12 Arnout Killian, *Foam*, 2002
3.13 Arnout Killian, *Frozen*, 2002
3.14 Glen Rubsamen, *The Call Of Cthulhu*, 2006
3.15 Glen Rubsamen, *Please Let My Affections Lead Me Into Danger*, 1999
3.16 Invitation card for opening *We'll slide down the surface of things...*, 2002

Bas Heijne: Real Seeing
4.1 Germaine Kruip, *2 Seconds*, installation view, Stedelijk Museum, 2000

Bregtje van der Haak: Interview with Tim Stoner
5.1 Francisco Goya, *'An heroic feat! With dead men!'*, plate 39 from the series 'Los Desastres de la Guerra', 1810–1820, © The Trustees of the British Museum
5.2 Francisco Goya, *Another madness of his in the same ring*, plate 19 from the series 'Tauromaquia', 1815, © The Trustees of the British Museum
– Thomas More, *Utopia*, book cover, originally published 1516
5.3 Bregtje van der Haak, *De vrijetijdszone*, film still, 2001, © DNW/VPRO
– J. G. Ballard, *Cocaine Nights*, book cover, 1996
5.4-5.5 Bregtje van der Haak, *De vrijetijdszone*, film still, 2001, © DNW/VPRO
5.6 Tim Stoner, *Long Haul*, 2000
5.7-5.8 *The Leisure Society*, Tim Stoner, installation view, De Vleeshal, February–March 2001, photo: Leo van Kampen
5.9 Tim Stoner, *Sierra*, 2000
5.10 Tim Stoner, *Development*, 2000

Valentijn Bijvanck & Rutger Wolfson: Museums and Contemporary Language. Interview with Geert Mul
6.1 Geert Mul, *then and now*, 1990, courtesy the artist
6.2 Geert Mul & Leon Dekker, *Virtual Expo*, 1994, courtesy the artist
6.3-6.12 Geert Mul, *Lowlands VJ set*, 1999, courtesy the artist
6.13 Geert Mul, *Generating Live*, video still, 2000, courtesy the artist
6.14-6.19 Geert Mul, *Generating Live*, installation view, De Vleeshal and De Kabinetten van De Vleeshal, August–September 2000, photo: Leo van Kampen
6.20-6.26 Geert Mul, *Tokyo FX*, video still, 1995, courtesy the artist
6.27-6.28 Geert Mul, *Babel*, installation views, 1995, courtesy the artist
6.29 Geert Mul, *This Land is Man-Made*, installation view, 2000, courtesy the artist

Chris Darke: Sublime Bodies. On the Work of Chris Cunningham
7.1 Chris Cunningham, *Come to Daddy*, album cover artwork, Warp Records, 1997
7.2 Chris Cunningham, *flex*, film still, 2001, Anthony d'Offay Gallery, London
7.3 Chris Cunningham, *Windowlicker*, music video stills, 1999, Warp Records, London
7.4 Chris Cunningham, *Come to Daddy*, music video stills, 1997, Warp Records, London
7.5 Chris Cunningham, *flex*, film still, 2001, Anthony d'Offay Gallery, London
7.6-7.8 Chris Cunningham, *Monkey Drummer*, video stills, 2001, Anthony d'Offay Gallery, London
7.9 Chris Cunningham, *flex*, film still, 2001, Anthony d'Offay Gallery, London
7.10-7.11 Chris Cunningham, *All is Full of Love*, music video stills, 1999, One Little Indian, London
7.12-7.14 Chris Cunningham, *flex*, film stills, 2001, Anthony d'Offay Gallery, London
7.15 Chris Cunningham, *flex*, installation view, De Vleeshal, March–May 2001, courtesy Anthony d'Offay Gallery, London, photo: Leo van Kampen

Edwin Carels: Unguarded Moments
8.1 *Terug van School*, installation view, De Vleeshal and De Kabinetten van De Vleeshal, September–October 2000, photo: Leo van Kampen
8.2 Claes Oldenburg, *Soft Alphabet*, 1978, Museum Boijmans Van Beuningen, Rotterdam
8.3-8.4 *Terug van School*, installation views, De Vleeshal and De Kabinetten van De Vleeshal, September–October 2000, photo: Leo van Kampen
8.5 Maison Martin Margiela, Remake of a cavalry cape, Fall/Winter collection 1995–1996, Museum Boijmans Van Beuningen, Rotterdam
8.6 Duane Hanson, *Seated Child*, 1974, Museum Boijmans Van Beuningen, Rotterdam
8.7 Patrick Van Caeckenbergh, *Collection de peaux*, 1991–1992, © Frac Champagne-Ardenne, Rheims

8.8 Luc Compernol, *Covering #1: Beautiful*, 1999
8.9 Television footage of Marc Dutroux's decent from the Neufchâteau steps, 1998
8.10 Marc Dutroux in Africa in the 1950's, © Photo News
8.11 Marijke van Warmerdam, *Handstand*, 1992, Collectie MuHKA, Antwerp, photo: Leo van Kampen
8.12 Robert Gober, *Untitled Shoe*, 1990, red casting wax, 7.6 x 7.6 x 19.7 cm, edition of 35 + 6 AP, photo: Adam Reich
8.13 Sharon Lockhart, *Audition Two: Darija and Daniel*, 1994, Museum Boijmans Van Beuningen, Rotterdam
– Dan Graham, *Rock my Religion. Writings and Projects 1965-1990*, book cover, Brian Gallis (ed.), MIT Press, Cambridge, MA, 2004
8.14 Walter Van Beirendonck and Paul Boudens, *Mutilate*, book cover, 1998, © Walter Van Beirendonck/Imschoot, Ghent
8.15 Inez van Lamsweerde, *Thank you Thighmaster: Pam*, 1993, Museum Boijmans Van Beuningen, Rotterdam
8.16 Wim Delvoye, *Sybille II*, 1999
8.17-8.18 Elke Boon, *Scrub*, 1997
8.19-8.21 Linda Dement, *In my Gash*, 1999
8.22 Günther Förg, *Chambres d'Amis*, 1986, collection S.M.A.K., Ghent, photo: Dirk Pauwels

Rutger Wolfson: This is the Flow – the Museum as a Space for Ideas
Formerly published in Rutger Wolfson, *Het museum als plek voor ideeën, essay*, Valiz, Amsterdam 2007

Anna Tilroe: The Promise
'De belofte, Ana Maria Tavares', formerly published in the series 'Vergezichten', in *NRC Handelsblad* and in Anna Tilroe, *Het blinkende stof. Op zoek naar een nieuw visioen*, Querido, Amsterdam 2002

Bas Heijne: Reality
'De werkelijkheid', formerly published in *NRC Handelsblad* and in Bas Heijne. *De werkelijkheid*, Prometheus, Amsterdam 2004

Cornel Bierens: We'll slide down
Formerly published in *NRC Handelsblad*

Bas Heijne: Real Seeing
'Echt zien, Kunst rond de N470', formerly published for the Province of Zuid-Holland (South-Holland)

Bregtje van der Haak: Interview Tim Stoner
Formerly published as part of the invitation for the exhibition The Leisure Society, De Vleeshal, Middelburg 2001

Rutger Wolfson: Art in Crisis
'Kunst in crisis', formerly published in Rutger Wolfson (red.), *Kunst in Crisis*, Prometheus, Amsterdam 2003

Valentijn Byvanck & Rutger Wolfson: Museums and Contemporary Language, Interview with Geert Mul
'Musea en de taal van het heden', formerly published in Rutger Wolfson (red.) *Kunst in Crisis*, Prometheus, Amsterdam 2003

Anna Tilroe: The Better World
'Het oceanisch verlangen, Higher Truth No. 5', formerly published in the series 'Vergezichten', in *NRC Handelsblad* and in Anna Tilroe, *Het blinkende stof. Op zoek naar een nieuw visioen*, Querido, Amsterdam 2002

Guus Beumer: The Language of Fashion
'De taal van de mode', formerly published in Rutger Wolfson (red.) *Kunst in Crisis*, Prometheus, Amsterdam 2003 and in *Morf*, no. 1, Amsterdam 2004. Parts of this article were previously published in *Forum Magazine* ('Around the Hearth') and included in a lecture on Democratic Design delivered at Casco in Utrecht.

Chris Darke: Sublime Bodies, On the Work of Chris Cunningham
Written for this publication

Edwin Carels: Unguarded Moments
Formerly published as essay for the exhibition *Terug van school*, De Vleeshal, Middelburg 2000

Anna Tilroe: On Values and Symbols
Bregtje van der Haak, Chris Keulemans, Jos de Putter, Marjolein Rothman, Anna Tilroe, Annelys de Vet, Cor Wagenaar and Rutger Wolfson – and in the second conversation, Alex van de Beld as well – took part in two discussions at Hotel De Filosoof in Amsterdam on whether there is a need for artists to develop new symbols, and what values such symbols should express. These conversations preceded the exhibition *New Symbols for the Netherlands* at De Vleeshal in Middelburg and were originally published in the essay collection of the same name edited by Rutger Wolfson and published by Valiz, Amsterdam, 2005.

Editor: Rutger Wolfson
Authors: Guus Beumer, Cornel Bierens,
Valentijn Byvanck, Edwin Carels,
Chris Darke, Bregtje van der Haak,
Bas Heijne, Anna Tilroe, Rutger Wolfson
Coordination: Astrid Vorstermans
Assistance: Jacqueline Cijsouw,
Angela Verschelling
Translation: Pierre Bouvier, Gerard Forde,
Anna Wolfson
Copy-editing: Pierre Bouvier, Gerard Forde
(introduction), Els Brinkman
Image research: Erik Tijman
Graphic design: Mevis & Van Deursen with
David Bennewith, Amsterdam
Lithography and printing: Bariet, Ruinen
Binding: De Haan, Zwolle
Publisher: Valiz, Amsterdam *www.valiz.nl*

ISBN-13: 978 90 78088 24 0
NUR 646, 651
Printed and bound in the Netherlands

valiz